Lecture Notes
in Control and Information Sciences   231

Editor: M. Thoma

Springer-Verlag London Ltd.

S.V. Emel'yanov, I.A. Burovoi and F. Yu Levada

Control of Indefinite Nonlinear Dynamic Systems

Induced Internal Feedback

Translated from the Russian by P.S. Ivanov

Springer

Authors

Professor Stanislav V. Emel'yanov
The Academy of Sciences of Russia, Institute for System Analysis,
9, Prospect 60 let Octyabrya, Moscow 117312, Russia

Professor Isaak Burovoi
Moscow Steel and Alloys Institute, 4, Leninsky Prospect, Moscow 117936, Russia

Dr Fedor Levada

British Library Cataloguing in Publication Data
Emel'yanov, S. V.
  Control of indefinite nonlinear dynamic systems : induced
  internal feedback
  1.Nonlinear control theory   2.Feedback control systems -
  Dynamics
  I.Title   II. Burovoi, I.A.   III.Levada, F. Yu
  629.8'36

  ISBN 978-3-540-76245-4   ISBN 978-3-540-40905-2 (eBook)
  DOI 10.1007/978-3-540-40905-2

Library of Congress Cataloging-in-Publication Data
A catalog record for this book is available from the Library of Congress

Typesetting by Editorial URSS, tel. 007 (095) 135-4246, e-mail urss@domar.pvt.msu.su

69/3830-543210 Printed on acid-free paper

Contents

Contents ix

Preface

This book is written for a wide circle of readers who take interest in concepts of basic sciences, specifically in concepts of cybernetics, automation, and control. The text aims to provide a new technique for the solution of a wide class of automatic control problems. The title of the course reflects the basic systematic approach the essence of which we will completely clarify in the third section. For now, we only note that the meaning of the word "induction" stems from the Latin word "inductio" which implies initiation, demonstration, stimulation, and so on. That is why mental associations arising in the minds of radio amateurs (the induced EMF in a conductor) and mathematicians (the induced metric in a vector space) are far from accidental. We will "induce" coordinates of a controllable object to obey a certain specific condition that is sufficient for ensuring the desired behavior of this object. The text considers simple examples to illustrate the basic problems, the most important aspects and terms of the theory of automatic control, and the essential difficulties that everyone has to do with when he/she faces the problems of automation of actual objects. The book describes comprehensively and in detail the most popular procedures of control synthesis, informally discusses their merits and shortcomings, states unresolved problems, and defines the urgent trends in the progress of the science of automatic control.

First of all, the book distinguishes the most complex and not yet fully investigated problem of control in the case of a deficiency of both the information on a controllable object and the characteristics of forces that act on it. The basic question under discussion lies in the following: is it always too necessary to know quite exactly the object of interest for its efficient automation or, perhaps, in a number of cases we can make do with the highly generalized information on its properties and characteristics?

In this connection, it is pertinent to note that the question raised above is in essence equivalent to the problem for developing a universal controller intended to control effectively a wide class of objects. It is also important to stress that a similar statement of the question contradicts to a certain extent the classical tradition of control theory by which the information on an object and external forces is always assumed to be complete.

What is likely to be the most striking fact is that the answer to the question put above proves positive in very many situations: there is no need to have the detailed information on the properties of an object for its high-quality control because it suffices to possess the information on its belonging to a fairly wide class of objects displaying some general or similar characteristics. Moreover, it turns out that under nonstrict constraints, the problem for the synthesis of the stabilizing control of a complex indefinite object (multidimensional, nonlinear, multiple-connected, and so on) can be reduced to a sequence of almost trivial operations. The specific character of this method is certainly defined by the decomposition rules which constitute not only the "zest" for the book, but also the dominant bulk of its subject matter. The basic idea of the principle successively developed in the book comes to the following: using local feedback loops, we strive to induce or, which is the same, to transfer to the system points, which are beyond the reach of a direct action, the desired control actions that must change the object dynamics in the requisite direction and eliminate the effects of uncertainty factors on this dynamics. It stands to reason that this approach leads to quite peculiar control structures which, however, exhibit a large "safety factor" and ensure a reliable compensation for interferences. In other words, it is the branching structures of controllers that impart them the properties of universality and flexibility and the capacity to control effectively a wide class of objects without the preliminary adjustment of their parameters.

It is worth noting that the above idea has a certain past history in the theory of automatic control. The idea that may most closely agree with the former one is the idea of the cascade or subordinate control already developed in the classical theory and used for the stabilization of power plants. However, a successful combination of the induced control concept and the principle of duality and generation of structures relating to the theory of control systems with new types of feedback loops has made it possible not only to extend appreciably the class of the objects automatized in this fashion, but also to produce essentially new and unexpected effects clearly illustrated in the book by examples of the engineering, physical, economic, and medical types.

Section 1 elucidates the basic notions and sets forth the key idea of the course, namely, the idea of the transfer of a control action through an intermediate system that acts as a controllable object subsystem. Practical examples are given in which this idea appears to be promising. Mathematical tools (ordinary differential equations) presented in the section serve to formalize the issues discussed in the lecture course.

Section 2 deals with the control problem for a special object which, in our opinion, is convenient for use as an intermediate system for conveying control actions. In this section, we design stabilizing control algorithms that are effective under the uncertainty conditions of object models.

Section 3 introduces the notion of an inducible internal feedback loop that corresponds to the mathematical implementation of the idea of conveying control actions through the intermediate subsystem of a controllable object. The technique is put forward for the solution of control problems. This technique involves the replacement of the problem under examination by a set of interrelated problems for inducing some loops. The synthesis of a control and the analysis of a closed-loop system are made for an individual induction problem.

Section 4 provides the procedure of designing control systems, which consists of a few interrelated induction algorithms.

Section 5 examines a few examples that relate to different fields of applications. In each of the examples, a pertinent induction algorithm solves a control problem.

In the Conclusion, we discuss trends toward further investigations of the issues touched on in the lecture course. We invite all readers to take part in these investigations and wish them every success.

It evidently makes sense to stress once again that this book is not a manual and does not contain prescriptions for the solution of specific problems. More likely, this is the book for interested readers, in which attempts have been made to outline some concepts in regard to possible ways of developing the basics of feedback theory and the synthesis methods of systems that display a requisite behavior. That is why we suggest this book to readers whose interest lies in the issues of the basic science and who are familiar with the basics of mathematical analysis and the theory of differential equations.

One of the authors of the book received the financial backing of the Russian Foundation for Fundamental Research and the European Economic Association, to which he is most grateful for the help.

We should like to express our appreciation to Prof. Ya. Z. Tsypkin whose constant attention and support were deciding factors in writing the book.

Prof. M. Thoma has played a specific role in preparing the manuscript. Although we met him not quite frequently, the discussions we held with him about the basic issues of the automatic control theory were quite useful for us.

The monograph could not have come out without the valuable discussions with and the concrete help of Prof. S. K. Korovin and Prof. Yu. S. Popkov to whom we express our deep gratitude.

We are grateful to P. S. Ivanov for his translation of the book into English and to Domingo Marín Ricoy for the preparation of Camera Ready Copy.

Authors Moscow, 1997

Control of Indefinite Nonlinear Dynamic Systems
Induced Internal Feedback

S. V. Emel'yanov, I. A. Burovoi, and F. Yu. Levada

The course covers 23 lectures and deals with a new technique for solving the problems of the synthesis of binary structures of nonlinear dynamic systems with induced internal feedback loops. The basic approach of this technique is to subject the coordinates of a controllable object to a definite condition so as to secure the desired behavior of the object as a whole. The general theoretical results outlined in the text are illustrated by the examples from biology, economics, power engineering, production engineering, and medicine. The course is intended for a wide circle of readers who take an interest in the problems of control and are familiar with the basics of higher mathematics.

Introduction

1. Processes, Systems, Actions: General Issues

In this lecture, we outline some problems, introduce the basic concepts, qualitatively describe the technique used for the solution of the problems, and discuss the attendant issues of the general nature.

1.1. What Do We Want?

It has long been noticed that some events perpetually occur in the environment. When it is said that some processes are running in the world, it is commonly kept in mind that particular characteristics of the world undergo changes with time: books are being written, galaxies are scattering, AIDS is spreading, the old are dying out, and the new is decaying — all of this can be interpreted as the course of certain processes.

What is inherent in people is that they are dissatisfied at how these processes are taking their course — even if not all the processes in the world, then, in any case, most of them. This dissatisfaction readily is being transformed to the desire of influencing them in one way or another to "improve the world". There arises a problem that this book will be dedicated to: it is necessary to exert such an action on the process of interest as to enable it to run in a definite desired fashion. Solutions of the problems of this type involve an appreciable portion of human endeavor.

1.2. What Can the Interrelation Between Processes Afford?

Interrelations between things and people in the world are obvious. This ancient formula is understood in the most different meanings. The workers of an enterprise are interdependent through production relations, the details of a work of art are interrelated by the common conception of the author, the future is linked with the past by cause-effect relations, the market prices of commodities depend on the competition involved, the planets revolving in their orbits are bound together by gravitational forces, etc. What will concern us here is the following meaning of this formula: the course of any process depends on that of any other processes (possibly, all of the remaining processes in the world).

The interrelations between processes increase our possibilities of "improving the world". A typical situation is that in which we have no possibility of acting directly on the process of interest. It happens sometimes that these possibilities exist in principle, but in view of some additional circumstances it would be undesirable to rely on them because they are disadvantageous in a certain sense. Then, it remains for us to resort to one approach: to produce an effect on the process that is accessible to actions, considering that this process will have its impact on the process of interest.

Thus, using the interrelation between processes, we can extend the sphere of our influence on the world and also make this influence subtler. It is this principle that political figures rely upon: it has long been known that the most convenient way to pull the burning chestnuts out of fire is to use somebody else hands. In essence, this text is written with the aim to extend the above approach to new fields of application, apart from the fields of politics and the thermal treatment of chestnuts, in which somebody else hands are used from time immemorial.

1.3. Language of the General Theory of Systems

For convenience of the further discussion, we will translate the statement of the problem in question and the suggested means to its solution into the language of the general theory of systems. In this language, the notion of a system (more precisely, a time-varying system) serves to formalize the intuitive concept of a certain device carrying out the process of interest: instead of the phrase "this is the way of how the process occurs", in this language we say "this is how the given system operates" or, which is the same, "this is the input of the given system". In the language of the theory of systems, an aggregate of factors that exert an influence on the course of the process under study is spoken of as an input action (the input of the given system). This system is said to convert such-and-such input to such-and-such output, keeping in mind that the run of the process depends on certain actions. The formal definition of such a system will be given below in Lecture 3.

In the language of the theory of systems, the problem for "improving the world" is stated as follows: it is necessary to influence the system of interest so that its output should get some definite properties. In this language, the interrelation between processes implies that the inputs for some systems serve as the outputs of other systems. We will use this interrelation to solve the stated problem as follows (Fig. 1.1). We expose a system to actions accessible to it so that the behavior of this system should have an effect on the system of interest in the desired manner.

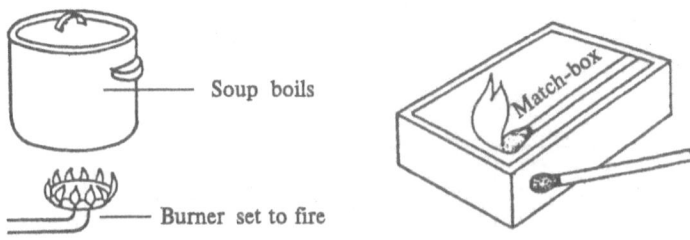

Fig. 1.1. Interrelation between processes at a kitchen. The intensity of boiling of soup in the pan (left) depends on the process of heat release from the burner. The other example (right) refers to the procedure of striking a match: the process of friction of a match head on the matchbox influences the process of heating of the match head. If in the first example the interrelated processes define the behavior of different physical objects (the burner is set to fire and the soup boils in the pan), in the second example we have to do with the only object — the match head.

1.4. A System and an Object

The treatment of a system as a device carrying out a certain process generates a need for examining such systems as those that do not agree with any objects of the material world. The way of defining such a system depends on what process we take interest in. In the simplest case, this can be the process of operation of a certain real object, and so the system coincides with this object. However, it makes sense to consider, for example, a system of heat transfer within the atmosphere or the immune system of an organism, although they do not exist in the world if they are taken separately. On the other hand, we can examine systems that comprise a set of independent physical objects. For example, if our interest lies in the dynamics of metal loss due to corrosion at the world scale, we should consider a system formed from all metallic objects, from the Eiffel Tower to the last nail in the heel of a Philippine police officer. Therefore, a system and a real object are different entities. In the text presented below we will sometimes use the term "object", as applied to certain systems, bearing in mind that the system at hand is the object under our examination or the controllable object (we will introduce this notion in Subsec. 3.4).

Note that the term "system" is often used to denote an "aggregate of interrelated elements" (for example, the system of ethical values of a society); in use are also other designations (for example, the railroad brake of the Westing-house system). Next, unless the contrary is stated, under the term system will be understood only a time-varying system — some entity the operation of which represents the process of our interest. In addition, the text will sometimes contain the phrase the "system of equations", but we hope that this will not put the reader to confusion (Fig. 1.2).

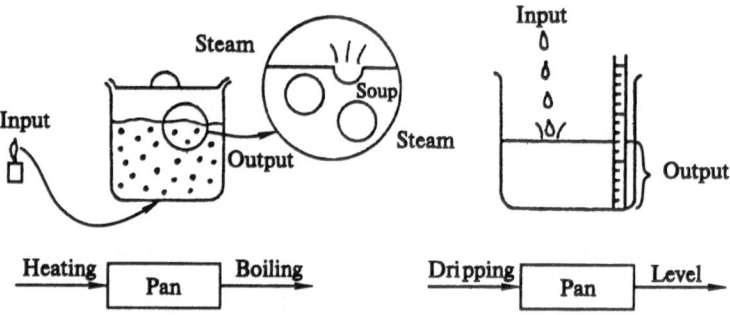

Fig. 1.2. The pan is a device that converts the process of heating of the pan bottom to the process of soup boiling (at top left). The requisite input-output conversion diagram is shown on the lower left. The theory of systems examines these conversions, disregarding all other features of appropriate objects. If the same pan is put under a leaky battery, this represents quite another system that transforms, for example, the process of dripping into the process of changes on the level of water in the pan (right).

1.5. Transfer of Actions Through the Chain of Systems

Using the modern language of the general theory of systems, we will describe the above method checked over centuries for the accumulation of the human power (in particular, the method of pulling the chestnuts out of fire). Let there exist two interrelated systems: the system A that we would like to set into oper-ation in the desired manner and the system B on which we can exert an action (stimulus). Assume that it is impossible to influence A directly. Then, having es-timated the effect of B on A, we can understand what action should be applied to B to enable B to set the system A to function in the desired manner. In this way, we can "reach" the inaccessible system A using B as an intermediate link in the chain of the transfer of our efforts. Let us note that these efforts can be subject to qualitative changes during their transfer: the actions of B on A and our actions on B can be different in nature (as a rule, this is just the case).

But then, why do we deal with only one intermediate link? Is it not possible to use two, three, or more links? Of course, it is possible. For example, by acting on a certain system C, we will exert an effect on the system B coupled to C.

The behavior of the system B will tell on the operation of the system A and will enable us to achieve the specified objective under certain conditions. Thus, using two intermediate links, we can reach the inaccessible system A. In a similar way, we can obtain a set of methods to act on A using for the purpose the chains of systems of different lengths. However, from a definite viewpoint, this wide range of the methods does not in principle contain anything new as against the case where only one intermediate link is used, namely, the system B. The whole point is how we interpret the system as a part of the world, which is responsible for the course of a certain process. In the case at hand, our interest lies in the fact that the chain of intermediate systems (links) takes up our action and carries out a process that exerts its influence on the inaccessible system A. So, we have the right to regard this entire chain as one system that responds to our action and exerts an effect on A; we considered this system earlier and called it the system B. A comparison with a postal system suggests itself: we send letters by post without displaying a particular interest in the sequence of transport operations performed for us by the postal service. On collecting all the links to form one system, we do not certainly have to forget their existence; In fact, we need to clarify how the system B operates in order to prepare a requisite stimulus. For this, it proves useful to know the design of the system B, although sometimes, as with post, the design of B has no meaning.

1.6. Direct Actions: A Small Trap

Thus, there is nothing radically new if the number of intermediate systems is more than one. But if it is less than one? What is a "direct action"? The question is not at all as simple as it appears at first glance. For example, the action produced by bare hands — is it a direct action? Perhaps, it is not because the hand can be thought of as an intermediate system that responds to our mental command, i. e., to the stimulus of the nervous system, and exerts an effect during its functioning on systems, namely, objects of the surroundings (the world around us). But the hand, too, cannot be acted upon "directly": between the hand and the brain there is an intermediate system — a nerve. Having continued the discussion in a similar way, it is easy to verify that "direct actions" do not exist in nature.

But any one measure intended to "improve the world" must include a "direct action" as initial impetus. But the latter is not possible, and so any "improvement of the world" is out of the question. In general, the human selfdependence is a pure illusion: all our actions are the actions determined by the world around us, and so the destiny of our consciousness is to contemplate passively this "puppet play", but then merely when there are impetuses to the contemplation.

In order not to take the bread out of the mouths of philosophers, we will not discuss the validity of the drawn conclusion. However, we cannot but acknowledge that the course of the conclusion is logically erroneous. The trap that we peeped in owes its appearance to the vague notion of a "direct action", which we attempted to use successively without previously taking the trouble to clarify it. We will make some amends for our negligence. An action can be considered as a process the effect of which on other processes is of interest to us. The action, just like any process, implies the functioning of a certain system. We will call this system the source of actions (the stimulus source). The behavior of a specific source of actions will be referred to as its direct action. Thus, a particular process is or is not a direct action depending on what kind of system is taken as the source of actions.

So, the above-described action taken by bare hands should be admitted as a direct action if these hands are constituents of the action source, in which case the operation of the source shows up just in a certain job done by the hands. Under definite conditions, we can consider the source of actions of this type to be a potter, a masseur, a participant of an election meeting, etc. Since our talk has turned to the hands, it is worthwhile to point out that even in the cases where a person performs a certain hand job, his hands do not necessarily exert a direct action on the world. First, a person can hold in his hands some tools such as a painter's brush, a surgeon's scalpel and a wood-cutter's axe. It is these tools that directly act on the world, whereas the hands are hidden somewhere inside the action source. Second, the hands can turn out to be beyond the source of actions: when a musician learns a new composition quite intricate for him, his disobedient hands play the role of the system A, namely, a portion of the world, which need be set to operate in the desired manner (Fig. 1.3).

1.7. Sources of Actions

What system can be taken as a source of actions (a stimulus source) depends on both the problem involved and the considerations of convenience. As an example, we will consider the following situation: a pianist plays the piano in a concert hall. If our interest centers on the neurophysiologic mechanism of motion of his fingers, we should acknowledge that the action source is the pianist without regard for his hands. If we consider the technique of the pianist's performance, the source of actions exerted on the piano is the musician, as such, together with his hands. If studies are made of the acoustic potentialities of the hall, the source of actions on the hall is the pianist with his hands and his piano. Finally, if our interest lies in the perception of sounds by listeners, the action source must include the pianist, his hands, the piano, and the entire hall.

The problem we are dealing with is to ensure a definite behavior of a certain system through the effect produced on the world by available means. Here, it makes sense to assume that the source of actions is a system of which we do

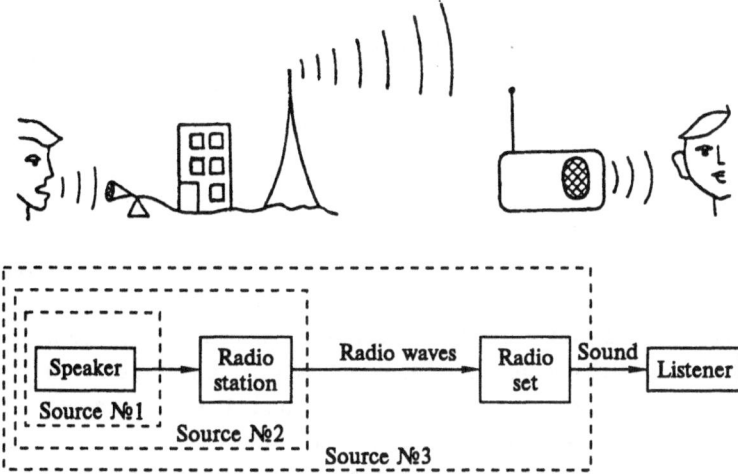

Fig. 1.3. Transmission of actions of a speaker's vocal cords on a listener's drum membranes occurs through the chain of intermediate systems. One version of this transmission is shown at the top of the figure. The diagram below offers a possible interpretations of this version in the language of the theory of systems. In this diagram, various methods in different cases can be used to represent an action source. The sources displayed in the figure are convenient, for example, for the description of the same situation involving different arrangements of shades of the meaning (in order of the numbers of sources, as displayed in the diagram): (1) President gave a radio report to people; (2) a capital radio station broadcast President's speech; (3) sounds of President's speech were heard from the dynamic loudspeaker.

not know beforehand that it will be such a function as we want. For example, the pianist knows that the sound waves in the concert hall atmosphere will carry to the listeners all shades of his performance, and so he has the right to regard the air oscillating at his own will as a source of actions. A person with a set of tools and a device can play the role of a source of actions, for example, a butcher with a knife, a society of cactus growers, and an automatic device such as the temperature controller of a refrigerator. Note that the behavior of these sources obeys the human will: the individual or collective will, or the will of the human being, which shows up in the design of a device. These factors reflect the features of the problem under examination: in the area of purely scientific exploration, natural "will-less" systems, too, such as volcanos and comets, can serve as sources of actions.

In the general case, the choice of action sources, which are optimal in a certain sense, for the "world improvement" is not simple at all, not to say hopeless. In practice, chaos reigns here, which is covered by the fig-leaf of the "historical definiteness": it is said that today these sources of actions exerted on the world are in use because yesterday the action sources were still worse. But if the action

sources had been chosen right from the start in another way, is it likely that the
world would be more perfect? And if the objectives, too, had been chosen in another way? Today, it is hardly probable that anyone can answer these questions.
We also leave these questions alone and enclose everywhere the phrase "world
improvement" in quotation marks.

Suppose that we have preassigned an action source and specified a system A
the operation of which need be put in correspondence with certain requirements.
If direct actions cannot permit us to achieve the objective pursued, it only remains
to find the system B linked with A, the former being accessible for direct actions,
and to attempt to act on A through B. Let us look ahead and see what happens if
it is possible to solve the problem. The system B will then exert on A that action
which ensures the desired functioning of the system A. Having learned to cause
B to behave in the desired fashion, we can incorporate this system into the action
source: this new source now acts on A directly. Thus, to carry out the transfer of
an action over an intermediate system requires the extension of an action source.
As it is known, the problem of this type was dealt with much time ago by our
legendary ancestor as he lengthened his arm with a stick. Therefore, the historical
roots of the approach under examination to the "world improvement" extend into
the past much deeper than the roots of the plot relating to burning chestnuts.

1.8. Is There a Need for an Intermediate System?

In the text presented above, we discussed the issue concerning the fact that it
is possible to combine the sequence of a few intermediate systems into one system
B (which is what we have done) and then to connect the input of the system to
the action source and the output to the input of the system A (on learning to
transmit over B everything that is necessary).

The question arises about where the system A has come from and why it lies
away from the intermediate system B.

The only certainly known thing is that the output of the system A represents
a process that we would like to act upon. The question is: why then our direct
action does not regard as the input of the system A? In this case, there is simply
no room left for the intermediate system.

This reasoning does not in principle lead to an error. Also, the design involving the two systems A and B also has the right to exist. Moreover, the design
comprising the two systems can serve as the methodological basis for the solution
of a number of problems for "improving the world".

The content of the course presented here just provides the foundation for
proving the latter statement. Thus, we give the affirmative answer to the raised
question as to whether the intermediate system is necessary. It will be clear from
the text that follows why we come to this conclusion.

2. Practice: Survey of Examples

The effect of transfer of actions via an intermediate system is not uncommon in practice: a number of the examples indicative of this effect are given in the text below. In all of the examples, two interrelated systems A and B appear, the latter being directly subject to our action. As regards practical goals, it is important that this action should exert a favorable influence on the behavior of the system A or cause the input itself of the system A to obey certain requirements, or secure the desired character of the inverse effect of A on the behavior of the system B.

2.1. Biology:
The "Predator-Victim" System

In the system of the "predator-victim" type, two populations of animals interact with each other — "predators" and "victims": it is only predators eating victims that can retard an increase in the number of victims, but an increase and a decrease in the number of predators depend on the amount of the available food, i.e., the number of victims. This system on its own generally undergoes fluctuations so that the population sizes of predators and victims vary in a periodic fashion. The thing is that the victims reproduced under favorable conditions enable the predators to multiply by virtue of the abundant grub, but the reproduced predators eat away with time an appreciable portion of the victims, so that they themselves die out of starvation. Fluctuations of the predator-victim system can actually lead to the predominance of agricultural pests or to the death of populations entered in the "Red book". It is hard to submit to this situation.

It is possible to influence the predator-victim system by varying artificially the size of one population or another, i.e., removing from the biocenosis a portion of species, or, on the contrary, releasing the animals bred in the artificial conditions. For example, biological methods of the protection of plants presuppose the artificial breeding and distribution of insects — "predators"; they retard the population size of herbivorous insects — "victims". Thus, the action on the population of predators (the system B) is transferred to the population of victims (the system A). In this situation, the cyberneticist could consider the question: how many incubation predators need be put out to live in nature so that the number of victims would be close to the desired level, for example, to zero?

Hunting-grounds can provide one more example of the action on the biocenosis. The hunt for grazing animals ("victims", i.e., the system B) impedes the growth of their population and hence the growth of the population of "predators" (the system A). If we remove too many victims from the biocenosis, both the victims and the starving predators can turns out to be on the verge of annihilation. But if we cease to hunt at all, fluctuations will emerge and both of the populations will be in distress from time to time. How should the hunt be controlled to avoid these troubles? This question is also worthy of notice.

2.2. Power Engineering: An Atomic Power Plant

The nuclear reactor of an atomic power plant converts the energy of nuclear decay to heat. To simplify the situation, it can be said that the reactor operation depends on two interrelated processes, the process of energy release and the process of heating and cooling.

Two systems correspond to these processes. Let the output of the system A be temperature and the output of the system B be the energy released per unit time, i.e., the power. It is possible to influence the behavior of the system by affecting the course of the nuclear chain reaction (the reactor design provides for requisite apparatus). In addition, the behavior of B depends on temperature, i.e., on the output of A, and the behavior of A depends on the power, i.e., on the output of B.

While acting on the reactor, we should ensure the desired behavior of the system B, i.e., obtain the desired power level. To achieve this objective, there is no need to use the effect of transfer of actions through the intermediate system because the system B can be acted upon directly. But this effect manifests itself without our wishes: acting on B, we exert an action on A, which will tell on the behavior of B. So, here we need to account for the effect of transfer of an action and, perhaps, impede its undesirable manifestations.

Let us note that in this example, one physical object — the reactor — corresponds to the two systems A and B, and it is in principle impossible to divide the reactor into portions that conform to these systems.

2.3. Economics: National Market Economy and State Budget

The behavior of the national economy can be defined by the processes of changes in two quantities: the fixed capital stock and the product output per unit time. Assume that these processes conform to the operation of the two systems A and B, respectively. We consider it desirable that the economy should develop so that the product output (i. e., the output of the system B) increases as fast as the technical progress and the job growth permit it to do so. Naturally, we deal only with those products which meet the demands of the consumer.

In the conditions of the free market economy, the operation of the systems A and B results from the joint activities of a number of firms, the behavior of which depends on the market condition. Thus, firms increase the output if the product marketing grows; as the output increases, the production expands and the fixed capital tends to rise. Hence, the output of the system B is the input of A and the input of B depends on the amount of marketing per unit time. A portion of the salable products is spent on an increase of the productive capacity and the other portion includes consumer goods. Therefore, the input of the system B can be taken as the output of A and the amount of consumption. The latter quantity involves both the commodity expenditure and services provided by the state, which follows from the appropriate budget items. So, the market situation is subject to the action of the state through the state budget. It is of interest to clarify what features this action must desplay to enable the economy to develop in the desired way.

Here we discover the same situation as that illustrated in the example from power engineering: it is necessary to ensure the desired behavior of the system B (commodity production) on which we can exert a direct action, although, this action arrives at the system A and then has its impact on B.

2.4. Production Engineering: Chemical Interaction Between Two Reagents

In metallurgy, heat engineering, chemical engineering, and other fields of industry, use is made of chemical reactions between two substances, which proceed with heat evolution. The operation of an appropriate set (a reactor) can be represented by three processes: the processes of changes in the amounts of the first and the second reagent and also the process of cooling and heating of the reactor mass. These processes are interdependent. The amount of any reagent within the reactor depends on the delivery from the outside of the reagent in the composition of the raw material, on the removal of unreacted residues in the waste products, and on the transformation of molecules of the reagent into some other molecules in the course of the chemical reaction. This transformation proceeds faster or slower depending on the temperature and the contents of the reagents. But the temperature depends on the reaction rate, i. e., the number of the above transformations per unit time, because each stage of the reaction results in the release of a certain amount of heat.

Assume that the delivery of one of the reagents to the reactor obeys our will. Let the system B specify a variation in the amount of this reagent in the reactor and the system B determines variations in two quantities — the amount of the second reagent and temperature. Here, the output of B serves as the input of A and the output A and our action serve as the input of B. Suppose there is a need to ensure the desired behavior of the system A. For example, we can imagine the operations performed on a certain burner with the aim to ensure the prescribed temperature and fairly complete burning of oxygen by way of changes in the amount of its delivery. For this, we need to establish in the appropriate manner the transfer of an action to A over the intermediate system B. It is sometimes necessary to subject the behavior of the system B, too, to certain conditions. Then, as in the examples taken from power engineering and economics, account should be taken of the inverse effect of A on B.

2.5. Medicine:
Hormonotherapy for Diabetes

For the life of a human being, it is necessary, among other things, that his blood should contain a definite amount of glucose dissolved in the blood. Glucose enters into the composition of food consumed by the organism and is used by cells as an energy carrier. For the organism to be able to function adequately, the content of glucose should lie within a definite range of values. In order that the glucose content should not be too low, a human being uses the mechanism intended to release and discharge into the blood the glucose stored in the organs and tissues of the organism; if the glucose store gets scanty, the human being needs to have a meal. For the glucose content to be not too high, there is another mechanism that ensures the deposition of glucose. This mechanism is set to work under the action of insulin — a pancreatic hormone.

We can reveal in the organism two interrelated systems A and B, the operation of which determines the processes of changes in the content of glucose and that of insulin. The output of system B serves as the input of A. Two physiological phenomena underlie the operation of the system B: the ejection of insulin from the pancreas into the blood and the autodecomposition of insulin in the blood. The efficiency of the pancreas depends on the glucose concentration, for which reason the output of the system A is the input of B. The interrelation between these systems in the organism of a healthy person affords an acceptable level of glucose in the blood.

The pancreas of a diabetic patient produces too small an amount of insulin or does not at all. In this case, the above hormone is injected artificially into the blood, which is often the only method of keeping the patient alive. The injection is an external action directly exerted on the system B which must then produce the desired effect on the behavior of the system A.

2.6. Mechanics:
Irregular Rectilinear Motion

Let us examine a material point (particle) capable of moving along a straight line under the action of a certain force the value and the direction of which can be preset and changed in an arbitrary way. Assume that the objective of this force action is to cause the material point to approach a certain target (a prescribed stationary point on the straight line) and then hold it near the target.

We introduce one-dimensional coordinates: let the quantity x express in certain units the lengths of the current deviation of the moving point from the target; the sign of x corresponds to the deviation direction (toward the right and left). The problem is to ensure an approximation of the current value of $x = x(t)$ to zero, where t is time. We consider the system A the operation of which shows up as a change in x with time, so that $x(t)$ is the output of the system. The motion of the material point, i. e., a change in x (t), depends on the speed $v(t)$ of the point at each instant of time: $\dot{x} = v(t)$. Here, the point above the symbol of the variable denotes the differentiation with respect to time in keeping with the Newton laws. In this connection, we will think of the quantity $v(t)$ as the input of the system A. Hence, the system A is such a "device" that converts v to x. The reader skilled in the mathematical analysis will readily find out an integral in this system. In the problem under consideration, the system A need be set to operate in the desired manner.

The behavior of the system A depends on a variable quantity — a process that does not represent our direct action. To exert an effect on the directly inaccessible system A, we should resort to a certain intermediate system B that must convert a force applied to the point to a speed of this point. This conversion has received the thorough study and relates to the second Newton law: $\dot{v} = F(t)/m$, where m is the mass of the material point at hand and $F(t)$ is the force applied to the point. Let us recall that $F(t)$ is our direct action on the system B.

What kind of output of the system B will exert the requisite effect on the system A and what kind of input will secure this output of B? This problem is given detailed treatment in a more common case in Lectures 7 and 8, and so we will not handle it here. Let us only note that the example considered in Subsec. 2.6 differs from all other examples in this lecture in the fact that the question raised in the example can be given an exact answer. It would be desirable to achieve a somewhat similar result in other areas of application.

3. Mathematical Tools and the Subject of Study

It is evident that we can give accurate specific answers only to the questions of the forms considered above. So far, words alone have served to clatify practical aspects; the questions related to these aspects can have only an indistinct metaphysical form, whereas under the guise of answers there can appear merely a more or less "profound" discussion on the subject of interest. The specific results presented in the example of mechanics owe their emergence to the fact that this example has been given, by virtue of the Newton laws, through the use of the available body of mathematics. As regards other areas of application, mathematics still takes a timid look at them. It would be desirable to open wider the door to it.

3.1. How Can We Turn from Words to Deeds?

What body of mathematics do we need to handle practical problems? We will select mathematical tools from the now known mathematical theories, without trying simultaneously to "improve" both the world and mathematics. Let us select out of these theories, such a theory as that which satisfies the character of the problems under examination to the highest degree. There is good reason to subject the selection of mathematical tools to the following rules. First, the selected tools must be able to operate on the basis of a mathematical language that permits us to express the results valuable from the practical viewpoint. In other words, the phrase of the type "if we act on the world so-and-so, it responds in such-and-such a way" must have a sufficienty clear practical meaning. Second, care should be taken of the fact that these mathematical tools afford the results on the basis of data which we can abstract in practice. Finally, third, the selected tools should be rather well evolved ones to enable us to derive nontrivial results: people have long persisted in "improving the world" and the time of many problems (the simplest ones) has passed.

Note that a mathematical tool in itself, being the means of introducing order among systems, can be taken as a system, too. The inputs of this system can serve to be data on the process of interest, the data being expressed in terms of the appropriate mathematical language, and the outputs can offer mathematical recommendations of "improving the world".

The requirements imposed on mathematical tools reduce to the following: the inputs of the tools (data on the world) should be readily obtainable, the outputs (mathematical results) should afford ease in their use, and the input-output conversion should occur not exactly in the obvious manner.

A problem for the choice a suitable branch of mathematics is free of a formal statement and hence has no formal solution. There is no reason to wonder at this fact. The mathematical tools in use represent constituents of a source of actions on the world; the choice of the source, as noted in Sec. 1, depends on camouflaged chaos (arbitrariness). We will also follow the path of arbitrariness, taking refuge, as is customary, in historical concepts. The result of our arbitrary choice is the body of mathematics involving the theory of differential equations.

3.2. Why Did We Turn to Differential Equations?

The theory of differential equations, which deals with the analysis of infinitesimals, appeared and developed as a special means for the investigation of real objects that differ in nature. There is no doubt that this theory serves as the most striking example of the answer of mathematics to specific practical questions of natural science. In the 18th and 19th centures, the analysis of nature and development of the theory of differential equations, occurred in parallel: the description of a certain process was taken to mean a pertinent equation. A belief arose that, in principle, everything was available for the conception of the world: it remained only to write out accurately all differential equations and investigate them. It would be erroneous to say that today not a trace remained from this selfconfidence. The traces do exist, and the main one refers to the stereotype technique of the analysis of real processes, in which under mathematization is virtually always understood the derivation of a differential equation.

One cannot but acknowledge that this technique suffers from its one-sidedness. But its merits are also selfevident. The long-standing predominance of the theory of differential equations in the area of mathematical methods for the analysis of nature has led to the fact that people comparatively well conceive how to abstract themselves from practice in terms of this theory and how to interpret theoretical results. Any other body of mathematics almost always proves to come off a loser; any theory, if it is spoken of in the language of systems, succumbs in that the inputs of a system are commonly more difficult to prepare and the outputs are more difficult to use. Thus, in many practical problems, the theory of differential equations has an advantage over other branches of mathematics in the two out of the three principles of the choice mentioned above, although this advantage is not of a principal theoretical character, but stems from the specific course in which the historical process of the science evolution took place. The third principle is a nontrivial one, which does not yet have a substantial significance in the field at hand: it is merely sufficient that the theory of differential equations sometimes enables us to obtain at least some mathematical results.

Thus, the theory of differential equations in regard to all of the three defined principles looks satisfactory and, without ruling out the possibility of the use of alternative mathematical methods, still retains the right to exist. What also supports this right is that over three centuries the differential equations have not managed to perform everything they are able to do. It is only when "old" mathematics exhausts its potentialities in all practical areas that the need will arise for completely switching to new branches of the mathematical theory. For the time being, there is still a very long way to do so.

3.3. Indeterminate Differential Equations

To understand how to force the system of interest to function in the desired manner, we perform the following. First, we describe the behavior of the system and the processes that have an effect on the system by means of differential equations. Here, we give a specific statement of the problem of interest and then examine the derived equations and study the influence of direct actions of the preassigned source on all of these processes. Finally, we use the result of these investigations so as to set up requirements for the operation of the source of actions.

The implementation of this procedure meets with grave difficulties even at the very begining. What is understood under the description of the process of interest by differential equations? For example, it is rightful to consider that the equations must adequately account for the dependence of the course of the process on all factors that influence it. But it is very difficult to allow for all the factors and essentially impossible to do so in many cases. For this reason, this insight into the described process often results in the abandoning of the attempts to use differential equations in practice and generally even any branch of mathematics. In all likelihood, it is this fact that is responsible for too low a level of mathematization of science.

We can avoid such an interpretation of the mathematical description of processes if we recall the objective for which the description is made up. The desired behavior of the system of interest is possible to achieve under our actions such as those which render the effect of a number of factors unessential. It can be said that if a nut serves as a fishing sinker, then the number of faces, the lead of thread, etc., are of no importance; what is only important is that the nut is a heavy piece with a hole suitable for fastening the fish line.

The description of the nut as a sinker with quite definite properties can include completely arbitrary features such as the types of faces and thread, and the manufacture. In short, different nuts can serve as sinkers to perform the same function. Similarly, in describing processes, there is no need to clarify all the factors that act on them. Here, the description of the process under study must include families of equations, one of which (no matter what kind of family) conforms to this process. Such a family of equations is commonly described by some representative of the general type, which can lawfully be referred to as an indeterminate differential equation. From the viewpoint of the general theory of systems, to the indeterminate equations there corresponds a system in which some of its inputs are unknown.

What are admissible limits of the indeterminacy of equations. The answer to this question determines how many facilities need be used to develop mathematical models of real processes. The answer itself can be found by investigating indeterminate equations of the requisite type and revealing the conditions under which the indeterminancy does not hinder us in finding the actions that produce the desired effect on the world. We will soon handle this question. Let us note that those activities in the "world improvement" which have to do with the investigations of mathematical abstractions, relate to the theory of automatic control.

3.4. Language of the Theory of Automatic Control

The theory of automatic control explores mathematical models of various processes and clarifies whether it is possible to influence these processes in the desired fashion, and if so, what actions we need to take for the purpose. The range of interests in this theory is not bound by any one specific practical field, be it production engineering, natural science, or humanitarian investigations. Technical objects are simpler than natural or social ones, for which reason there are more mathematical mootels designed for the former objects; this explains the origin of the legend as to the relation between the above theory and engineering. The language of the theory of automatic control gives a convenient way of describing the mathematical aspects destined for the "world improvement". In this language, a control is said to be a direct action exerted on the system of interest, a controllable object is a system the output of which need be put in correspondence with certain requirements, a controllable quantity is an output of the controllable object, and a control law (algorithm) is a mathematical model of the source of actions. In addition to the control action, the controllable object also has other input actions which are known as disturbances.

In terms of the theory of automatic control, the problem at hand shows up as follows: for the controllable object of interest there is a need to state a control law that secures the required character of changes in a controllable variable.

The problems of this type are known as the problems of automatic control. What distinguishes the class of automatic control problems, which we intend to consider in this text, is that a controllable object is defined by a system of ordinary differential equations which are indeterminate ones. Where they are taken from and what they mean will be clear once we represent the controllable object as a dynamic system.

3.5. What is a Dynamic System?

Dynamic systems belong to "input-output" systems, each being formally defined as a set of ordered pairs includig the elements of the sets of inputs and outputs. The first component of such a pair is viewed in the informative sense as an input action (input) and the second component as an output quantity (output).

It is commonly said that the system responds to an input action, generating an attendant value of the output quantity.

The conversion of an input action to a value of the output quantity sometimes displays the character of mapping (the function): the system responds uniquely to each action and is said to be a functional system. However, this simple behavior is not intrinsic to natural and technical objects. It is possible to reveal functional systems by resorting to idealization. An example of such a system is an idealized slot machine for sale of soda water: it responds to a coin that serves as an input action and releases 200 ml of syrup and water. However, as all know, real automatic devices behave in a much more complex way.

To describe the behavior of systems, which are not functional ones and in which different outputs can correspond to one input, we introduce the space of states or phase space. This is done in such a way that the output of the system is uniquely specified by its state (its phase) and by its input action. The choice of one state space or another generally depends only on the requirements for convenience of the description of the given system and is independent of the physical structure of the system.

The description of a system by means of a function that defines the system output as the response to the input action in the given state does not allow us to trace changes in the system with time (as the input-to-output conversion). To make up for this gap, we introduce state transition functions: the laws by which the system states undergo changes with time in response to input actions. If we prescribe an "input-output" system for which we select the state space, determine the responses to various input actions in all the states, and define the state transition functions, then it can be said that we have preassigned the dynamic system.

A state space for describing a system appears in Fig. 3.1.

3.6. Differential Equations as Models of Dynamic Systems

We will examine systems in which the inputs, outputs, and states are given as finite-dimensional vectors with real coordinates and the state transitions are described by integral equalities of the form

$$x(t) = x(t_0) + \int_{t_0}^{t} f(x, z, t)\, dt \qquad (3.1)$$

where $x(t) \in R^n$ is the system state at the time t; $z \in R^m$ is the input action; t_0 is the initial time; and $f : R^n \times R^m \times R \to R^n$ is a certain function. It is convenient to write (3.1) as the vector differential equation

$$\dot{x} = f(x, z, t) \qquad (3.2)$$

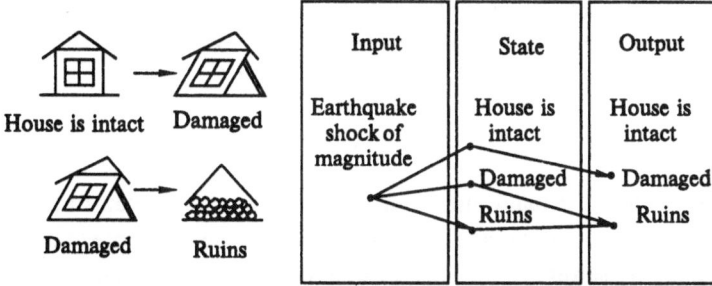

Fig. 3.1. The state space is used to describe a system capable of rearranging one and the same input to a few different outputs. For example, the state of preservation of a house (the output of the "house" system) after an input action such as an earthquake shock of a definite magnitude can vary depending on the initial house state that can also be specified by the degree of the house safety. If we assume for simplicity, that there are all three possible states (at the left of Fig. 3.1), then the response of the system to a definite input action will be such as desplayed in the diagram on the right, in which the transitions of states on exposure to input actions are shown by solid lines.

which must include the initial state x_0 of our system.

The function f in (3.2) enables us to reveal how the system state varies after the time t if the state at this time conforms to the point x and the system receives the input action x. Note that here, along with values of x and z, account should also be taken of the current value of the variable t for the mere fact that this value is one of the arguments of the function f. There is direct evidence of the contradiction with the common notion of the dynamic system: in Subsec. 3.5, the case in point only relates to state transitions under the influence of input actions, but not a word is said about such a variable as the current time.

We can remedy the situation by interpreting the variable t as one of the coordinates of the vector of input actions or as the state vector. In principle, these two ways are equivalent to each other.

According to the first way, the input of the system (3.2) is considered to be the pair of the variables z and t. In contrast to the input action z that can take any values and vary in the most intricate manner, the input variable t grows at a constant rate of $\dot{t} = 1$ and is always equal to the current time.

Following the second way, we assume that the pair x and t is a system state, the system (3.2) and $\dot{t} = 1$ describe state transitions with the initial point $x(t_0), t_0$. The vector z will represent the input as before.

We will take the second way, but allow ourselves quite a large freedom of discourse, i. e., we will describe state transitions by (3.2) and refer to z and x as the input and the state, respectively, although this reference was found to be improper. This freedom of discource is conventional in the theory of automatic control. The matter is that during investigations of dynamic systems, it often proves convenient to disregard initially the explicit dependence of f on t and to examine, instead of (3.2), an expression of the form

$$\dot{x} = f(x, z)$$

Here, the vector x actually turns out to be the state and retains this name in passing to a more complex case of the system (3.2). We will proceed in the same way recalling ourselves at times that this discourse is not too well.

As noted in Subsec. 3.5, the dynamic system model must describe in all the states the responses of the system to an input action as well as the state transition functions defined by (3.2). To put it differently, the model must also contain a function the values of which represent the outputs of the system, with the arguments serving as the inputs z and states x and t. In what follows, we restrict ourselves to dynamic systems in which the inputs are idential to states. There is no need to describe the response to input actions for these systems because their models are set by differential equations of the form (3.2) where the state and output are identical to each other.

3.7. Model of a Controllable Object

Let us combine components of the input vector z, which correspond to control actions exerted on the system (3.2), into the control vector u and shape the disturbance vector a from the rest of the components. We imply that a change in u obeys our will and a change in a corresponds to the effect of the world on the system under consideration.

We replace the input z by the pair of the vectors u and a and specify the model of the system by the equations

$$\dot{x} = f(x, a, u, t) \tag{3.3}$$

Sometimes, it makes sense to rest content with this notation of the model of a controllable object. However, one should not forget that the function f accounts for the nature of an object and its argument a reflects the nature of the environment, although both of these factors are far from always known precisely. In the majority of practical cases, the notation of (3.3) is indicative of nothing more than the unrealizable dream.

If the model (3.3) cannot be built up, one has to be content with indeterminate differential equations. Assume that the available information on f and a is sufficient to place the object of interest in correspondence with the family

$g : (x, u, t) \rightarrow x$ of mappings. On selecting arbitrarily a representative of this family, we can describe the object by an equation of the form

$$\dot{x} = g(x, u, t) \tag{3.4}$$

Note that in contrast to (3.3), the notation of (3.4) does not represent the model of a dynamic system, but only points to a family of equations (corresponding to functions g), amoung which the model is found. But for simplicity of the presentation, (3.4) will be spoken of further on as an actual model.

It should be noted that in practical problems it is common to supplement the differential equations for state transitions, even if indeterminate ones, with certain constraints on quantities or on the character of changes in these quantities of states and controls: the temperature (by the Kelvin scale) is nonnegative, the speed does not exceed the velocity of light, etc. We will not touch on this issue until after we take up specific examples presented in the last chapter of the lecture course.

3.8. Control Problems: How to Handle Them

In its general form, a control problem involves the estimation of a control u such that the point x moves in the desired manner on account of any equation of the family (3.4). Our main concern will lie in such a motion of the system that the variable x becomes close to a certain assigned vector variable; similar problems are known as tracking problems. We introduce limits on the method of shaping the current values of u and seek a control in the class of the functions $u = u(x, t)$. Convenience of the control of this type is due to the fact that the result of the substitution of $u(x, t)$ into (3.4), which represents a closed-loop system model, has the form

$$x = g(x, u(x, t), t)$$

and relates to the class of ordinary differential equations for the analysis of which some tools are available in mathematics. Thus, there is a hope of estimating the possibilities of the control law defined above.

To solve the control problem, we resort to the conceptions outlined in Sec. 1. In this case, we replace the system (3.4) by the pair of the systems A and B so that the control u directly acts on the system B and propagates its effect on A through B. As noted in Sec. 1, the choice of the systems A and B stem from the considerations of convenience and depends on properties of the family of the functions g in the model (3.4). The aim of our lecture course is to illustrate a definite approach to the solutions of different problems rather than to solve a definite control problem. For this reason, we will not describe the family of the functions g for which the method, still unknown to the reader, is suitable for use. Let us first concern ourselves with the presentation of the approach itself; the field of its application will show up on its own during our discussion.

We will begin with the mathematical implementation of direct actions on dynamic systems (this aspect is dealt with in the second chapter of the course). As a result, we obtain the description of a class of the systems B convenient in a certain sense for direct actions and a class of the pertinent control laws. Some portion of the control problems will be solved during the discussion.

Further, we should implement the transfer, as such, of an action through the system B (Chapter 3), investigate what action exactly the system A requires (Chapter 4), thereby completing the discussion of the approach, and, finally, consider examples in the last chapter.

Control
of Elementary Dynamic Systems

We define the systems A and B from the convenience considerations. The second system undergoes a direct action emerging from a certain source and transfers this action to the first system. What is meant by the term convenience? This chapter handles this question. We will specify a class of dynamic systems, which will then be included into the systems B, and suggest control algorithms for these systems. In this way, it will be "convenient" for us to single out just these types of B and apply to them the controls mentioned above. That is why we call the systems of this class elementary ones.

In passing, we will solve problems for which a controllable object belongs to the above class of the dynamic systems B; however, it is not worth overestimating the significance of these concurrent solutions. In addition, the results presented in this chapter will permit us to reveal a set of problems to which we will not apply the approach developed in this course: these are object control problems from which we cannot "cut off" the system B of the pertinent type.

4. Control of a One-Dimensional Object

The dimension of a controllable object is known as the dimension of its space of states, with each state conventionally defined in an ill-defined fashion. Therefore, in the strict sense, the systems considered below are two-dimensional ones if the current time is taken as a state coordinate.

4.1. Model of an Object and Control Capabilities

Consider the system

$$\dot{x} = \varphi(x, t) + u \qquad (4.1)$$

where $x \in R$, $u \in R$, and $\varphi : R^2 \to R$. We can represent the behavior of this system as a motion of the point x along a number line. How can this point be brought to zero? For this, we need to overcome the effect of the summand $\varphi(x, t)$ in the right side of (4.1), the summand being specified by the object nature and disturbances. If we apply the control to the system, this summand will obviously be "neutralized" and the system will not shift from the initial point. If we reinforce the control action by still another summand, it is this summand that will determine the character of motions of the closed-loop system. For example, if we put

$$u = -\varphi(x, t) - kx \qquad (4.2)$$

where $k = $ const, the solution of the system (4.1), (4.2) assumes the form

$$x(t) = x(t_0)e^{-k(t-t_0)}$$

At fairly large values of $k > 0$, the exponential function decays rapidly; we take it that this character of solutions conforms to the statement of the problem.

The system consisting of the controllable object (4.1) and control (4.2) is said to be the close-loop system (or the closed system, for short) of automatic control.

The control law (4.2) is applicable to a closed system only if the function $\varphi(x, t)$ is known. This is a very strong requirement and is not commonly met in practice. Therefore, the control (4.2) cannot be accepted as satisfactory.

4.2. What Can a Majorant Do?

In the system (4.1), (4.2), the signs of the quantities $\dot{x}$ and x are inverse, and so the system moves to zero; in this case, if the quantity $\dot{x}$ is rather large in magnitude, the system moves to zero fairly fast.

These results are obtainable without the use of information on values of $\varphi(x, t)$. Indeed, if

$$\operatorname{sgn} u = -\operatorname{sgn} x, \quad |u| > |\varphi(x, t)|$$

then the closed system moves to zero. For this motion to proceed quite fast, it still remains for us to ensure that a value of $\varphi(x, t)$ exceeds a value of u to a definite extent.

To implement the suggested concept, we introduce a function such that at all x and t we have

$$\Phi(x, t) \geqslant |\varphi(x, t)| \qquad (4.3)$$

Here $\Phi(x, t)$ is a majorant of the absolute value (modulus) of $\varphi(x, t)$.

As a rule, real objects afford sufficient data for estimating this majorant. For example, if the function $\varphi(x, t)$ can be given as the polynomial

$$\varphi = a_0(t) + a_1(t)x + \ldots + a_m(t)x^m$$

in which the coefficients vary in an unknown manner within the range

$$|a_i(t)| \leqslant A, \quad i = 0, 1, \ldots, m$$

then the majorant is found to be the function

$$\Phi = A(1 + |x| + \ldots + |x|^m)$$

Generally speaking, the majorant is sought in a specific way for each particular problem; the universal methods of majorization are nonexistent.

Assume that we have found the function Φ satisfying the condition (4.3). Obviously, when $k > 1$, the control

$$u = -K\Phi(x, t)\,\mathrm{sgn}\,x \tag{4.4}$$

ensures that a solution of the system tends to zero everywhere, excepting those points at which $\Phi(x, t) = 0$. In order that the system should not stop anywhere other than at zero, we subject the majorant to an additional condition (the majorant is totally in our power): $\Phi(x, t) > 0$ when $x \neq 0$. In this case, any solution of the closed system always approaches zero.

It is still early to finish the discussion, assuming that we have solved the problem. So far, the main question has not received due attention: does a solution of the closed system exist at all? In fact, for a nonexistent object, any assurtion as regards its properties will be true. Is it possible that our conclusion on the character of motion of the point x belongs to the truth of this type?

4.3. Problem of Correctness and Completion of Control Synthesis

Before proceeding to the discussion of the properties of trajectories of a closed system, we need to be sure that these trajectories exist. For this, it is sufficient to verify that the conditions of the classical theorem of exisistence and uniqueness are valid for the solutions of ordinary differential equations. It is only this way that we apply in our further discussion to all controllable systems built up according to the approach described.

Recall that for the vector equation

$$\dot{x} = g(x, t)$$

these conditions consist in the following:

(1) the function g is continuous;

(2) the function g satisfies the Lipschitz condition with respect to x on any bounded set (for brevity, we will say in this case that the function g is a local Lipschitz one in x) (Fig. 4.1).

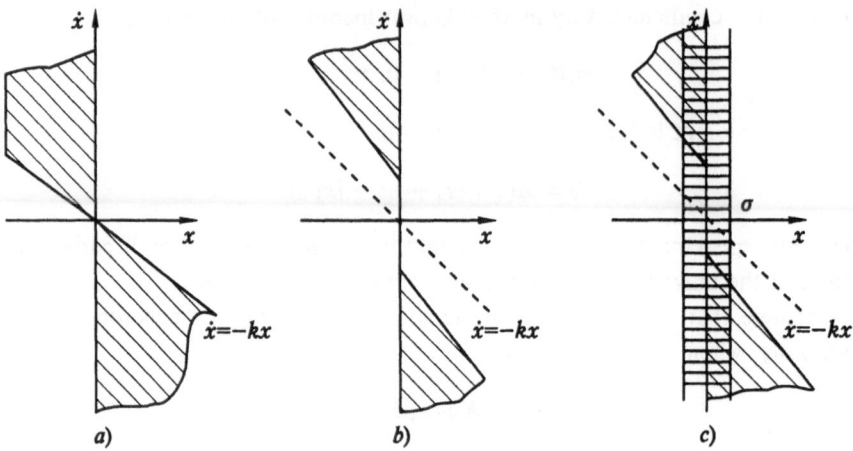

Fig. 4.1. Opposite signs of the speed and the coordinate of the system ensure that values of x tend to zero. A more severe bound on the pair of numbers of x, $\dot{x}$, which corresponds to the shaded region of Fig. 4.1 a, leads to the estimate $x(t) = O(e^{-kt})$ that can be taken satisfactory at an appropriate value of k. The contribution of the control to the speed must afford this bound under the uncertainty conditions of the system model. For some versions of the right portion of the model, it is necessary to secure a more severe bound, sometimes such as displayed in Fig. 4.1 b. The discontinuity of the quantity $\dot{x}$ in this case makes the system generally unrealizable. To obviate this undesirable circumstance, we can admit arbitrary values of x in a narrow band $|x| \leqslant \sigma$, as depicted in Fig. 4.1 c. In such a system, a value of x approaches zero until $|x| \geqslant \sigma$ and then it leaves the σ-neighborhood of zero.

Here, we find a unique solution for any initial point, although this solution is not liable to exist all the time; the point can move to infinity over a finite time period.

Let us clarify whether the conditions of this theorem have been met for the equations of the system (4.1), (4.4). In what follows, we will assume everywhere that the model of the system with the zero control is correct, i. e., it satisfies the above conditions. In the case under consideration, it is then sufficient to ascertain that the control action (4.4) is continuous and displays the local Lipschitz property with respect to x.

The cofactor sgn x in the law (4.4) is discontinuous at the point $x = 0$ and hence the control can be continuous only if $\Phi(0, t) \equiv 0$. This identity does not contradict inequality (4.3), but only if $\varphi(0, t) \equiv 0$, i. e., if the model does not contain a free term in its right side. If, in addition, the function is continuous and the local Lipschitz one in x, then the control (4.4) secures the correctness of the model of the closed system. In this case, the solution $x(t)$ exists and displays the same properties as those discussed above.

If a free term in the right side of the model is available, the situation becomes more complex. Of course, if values of the function $\varphi(0, t)$ are accessible for measurements, we can completely offset the function by a control action, thus reducing the problem to the one considered earlier. Indeed, let us rewrite the system model in the form

$$\dot{x} = \varphi(x, t) - \varphi(0, t) + [\varphi(0, t) + u]$$

and assume that the expression placed in square brackets represents a control.

Of interest is the case where the function $\varphi(0, t)$ is unknown and we are led to apply everywhere the positive function $\Phi(x, t)$. Here, the control (4.4) is discontinuous at zero and is inapplicable in this case. To avoid the discontinuity, we introduce into the control law such a cofactor as that which goes to zero at the point $x = 0$. Near this point, the values of the cofactor are bound to be close to zero from the continuity considerations. Hence, at low values of x, the control action will be low in magnitude and, in general, will not be able to convey the desired character of motions to the object. We will assume that this is the pay for continuity.

We prescribe a constant $\sigma > 0$ and determine both the continuous function and the local Lipschitz function $\widetilde{\Psi}(|x|, \sigma)$ in x so as to fulfill the conditions

$$\widetilde{\Psi}(|x|, \sigma) \geqslant 1 \quad \text{for } |x| \geqslant \sigma, \ \widetilde{\Psi}(0, \sigma) = 0$$

A set of functions such as

$$\widetilde{\Psi} = \min\left\{1, \frac{|x|}{\sigma}\right\}, \quad \widetilde{\Psi} = \frac{2|x|}{|x| + \sigma}$$

satisfy these conditions.

Let us examine a control law of the form

$$u = -k\Phi(x, t)\widetilde{\Psi}(|x|, \sigma)\,\mathrm{sgn}\,x \tag{4.5}$$

If the function $\Phi(x, t)$ is continuous and the local Lipschitz one in x, then the model of the closed system (4.1), (4.5) is correct. Beyond the σ-neighborhood of zero, the control (4.5) acts in the same way as (4.4) and enables the system to approach zero. Within this neighborhood, any character of motions can be accepted as being satisfactory if a value of σ is rather small and if the system cannot go beyond this neighborhood. It is evident that in our case, both of these conditions can be met.

The number of the adopted assumptions is not yet sufficient to make sure that x tends to zero when $\varphi(0, t) \equiv 0$ or, otherwise, that x falls within the σ-neighborhood of zero in a finite time. These issues are dealt with in the text below.

4.4. Character of Transient Processes

A closed system can be defined by the equation

$$\dot{x} = \varphi(x, t) - k\Phi(x, t)\widetilde{\Psi}(|x|, \sigma)\,\mathrm{sgn}\,x \qquad (4.6)$$

where $\widetilde{\Psi} \geqslant 0$ at $|x| \geqslant \sigma > 0$ and $\widetilde{\Psi} = 0$ at $x = 0$ and $\sigma = 0$. The equality $= 0$ is admissible only if $\Phi(0, t) = 0$. Reasoning from the fact that Φ is the majorant of $|\varphi|$ and considering the constraints imposed on $\widetilde{\Psi}$, we obtain the following equality at $|x| \geqslant \sigma$ (here, the quantity σ can be equal to zero):

$$\frac{d}{dt}|x| = \dot{x}\,\mathrm{sgn}\,x = \varphi\,\mathrm{sgn}\,x - k\Phi\widetilde{\Psi} \leqslant -(k-1)\Phi \qquad (4.7)$$

Therefore, when $k > 1$, the quantity x strictly decreases in magnitude throughout when $\Phi(x, t) > 0$. According to the accepted assumptions, this inequality is always valid when $x \neq 0$. Therefore, the point x approaches either zero or the σ-neighborhood of zero if $\sigma > 0$.

However, this does not yet mean that $x > 0$ or that the point reaches the σ-neighborhood of zero. The thing is that the function $\Phi(x, t)$ can rapidly decrease with time at each value of x (but it always remains positive). This can lead to a retarded motion of the system. Let, for example,

$$|\varphi(x, t)| = \Phi(x, t) = e^{-t},\ \sigma > 0,\ \widetilde{\Psi}(|x|, \sigma) = 1, \quad \text{when } |x| \geqslant \sigma$$

Assume that at the initial time $t_0 = 0$, the system is found to be at the point $x_0 > \sigma$. In view of (4.6) we have $\dot{x} = -(k-1)e^{-t}$. The function presented below is the solution of the above equation:

$$x(t) = x_0 + (k - 1)(e^{-t} - 1).$$

Therefore, $x(t) \to x_0 - (k - 1)$ and if $x_0 > k - 1 + \sigma$, the system does not reach the σ-neighborhood of zero because it cannot pass through the point $x_0 - k + 1$.

To rule out the situation of this type, we will require that the function Φ should take rather high values at all t. Let

$$\Phi(x, t) \geqslant M|x| \qquad (4.8)$$

where M is a positive number. Then, in view of (4.7) at $|x| \geqslant \sigma$, we get

$$\frac{d}{dt}|x| \leqslant -(k-1)M|x|$$

and hence

$$|x(t)| \leqslant |x(t_0)|e^{-(k-1)M(t-t_0)} \qquad (4.9)$$

It is now clear that the point x in the closed system either tends to zero or reaches the σ-neighborhood of zero in a finite time.

According to (4.9), the following asymptotic formula is valid for the solution of the closed system:

$$x(t) = O(\exp(-(k-1)M(t-t_0))) \qquad (4.10)$$

where the constant O depends on the initial conditions[1].

We will suppose that the exponential asymptotics of the decrease satisfies the requirements of the control problem. Hence, selecting a fairly high value of $k > 1$ and/or $M > 0$, i.e., taking $\Phi(x, t)$ such that tha constant M in (4.8) has a comparatively high value, we can ensure the desired character of motion of the closed system.

Diagrams illustrative of the continuity of control at zero appear in Fig. 4.4.

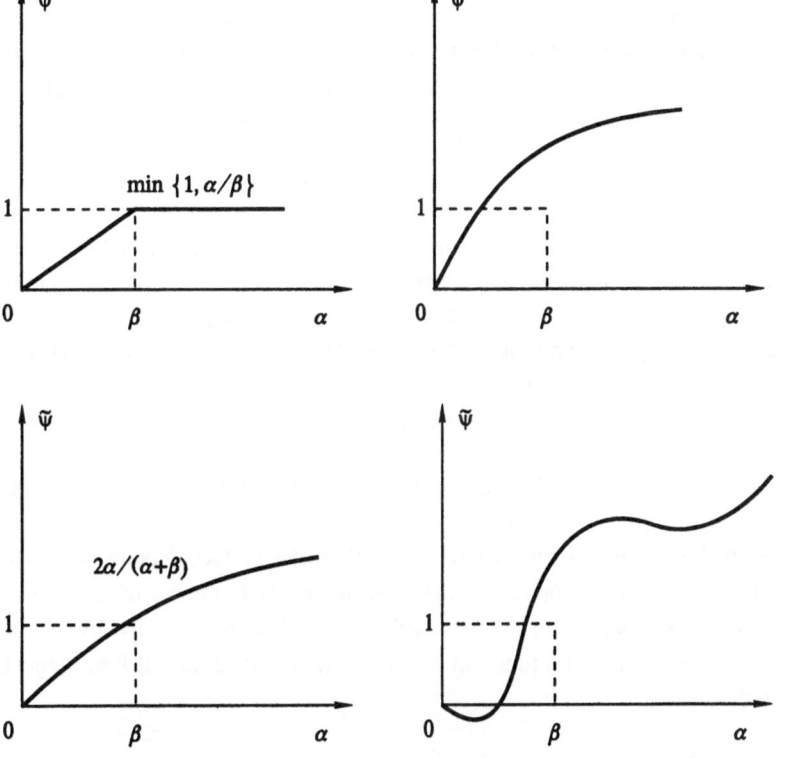

Fig. 4.2. Continuity of control at zero is achieved through the introduction of the smoothing cofactor $\widetilde{\Psi}(|x|, \sigma)$. The function $\widetilde{\Psi}(\alpha, \beta)$ must satisfy the conditions $\widetilde{\Psi}(0, \beta) \equiv 0$ and $\widetilde{\Psi}(\alpha, \beta) \equiv 1$ at $\alpha = \beta$. Examples of the functions of this type are given in the figure.

[1] Recall that by definition, we get $f(\xi) = O(g(\xi)) \Leftrightarrow \exists c > 0 : |f(\xi)| \leqslant c|g(\xi)|$.

4.5. Extension of the Problem: An Indeterminate Coefficient Involved with a Control Action

As seen from the above discussion, a closed system behaves in the desired fashion if, first, a control action is fairly high in the absolute value and, second, has the sign that is the inverse of the sign of the signal x. Evidently, the system of a more common form

$$\dot{x} = \varphi(x, t) + b(t)u \qquad (4.11)$$

will display the required character of motions if it is possible to impart the two properties mentioned above to the product $b(t)u$. We will clarify the situation in which there is a possibility to do so.

The coefficient b must obviously not go to zero. If its value is known at each instant of time, the replacement of variables permits us to reduce the problem to the solved one.

We consider the case where values of $b(t)$ are inaccessible for measurements. Here, the product bu can be taken rather high in magnitude if a value of $|b|$ is always comparatively high; the required sign of bu is secured if the sign of b is known. Consequently, we can assume that the following condition is valid:

$$0 < \frac{1}{B} \leqslant b(t) \leqslant B \qquad (4.12)$$

where $B > 0$ is a certain known constant. Strictly speaking, the right end of this range has no value; here we indicate the value at this end to satisfy the conditions of practice for which the unlimited parameters are not typical.

It can readily be seen that for the object (4.11) with the constraint (4.12) we can use the control law derived previously, therewith reinforcing it by the constraint $K > B$. The solution of the closed system will then have the asymptotics

$$x(t) = O\left(\exp\left(-\left(\frac{K}{B} - 1\right)M(t - t_0)\right)\right) \qquad (4.13)$$

Note that an unknown current value of b can in fact depend not only on time, but also on x, u, or some other measurable or nonmeasurable arguments. But this is of no significance if the condition (4.12) holds. In (4.11) we indicate only one argument of the function b, which, however, does not limit generality of the discussion.

4.6. Extension of the Problem: Additional State Parameters

Assume that the controllable quantity x is a coordinate from the set of coordinates of the state vector for a certain multidimensional dynamic object. Let this coordinate obey the equation

$$\dot{x} = \varphi(x, \xi, t) + b(t)u \tag{4.14}$$

where ξ is the vector of the state parameters added to x. There is no need to control these parameters. Suppose that the condition (4.12) is valid and the function $\Phi(x, \xi, t)$, which is the majorant of the quantity $|\varphi|$, is known.

If $\Phi(0, \xi, t) = 0$ at all ξ and t or $\Phi(x, \xi, t) > 0$ at all x, ξ, and t, then the problem does not in essence differ from that we have solved. Of interest is the case where $\Phi(0, 0, t) > 0$, but $\Phi(0, \xi, t) > 0$ at certain values of ξ and t. Here, we can obviously prescribe a constant $\sigma > 0$ and ensure that the coordinate x falls within the σ-neighborhood of zero. But, recall that a positive value of σ determines the payment for the continuous control which in this case does not involve any charge. At $\sigma > 0$, there is no way of causing x to tend to zero if the vector components vanish. To retain this opportunity, one should learn to use such a σ-neighborhood of zero that its radius goes to zero at $\xi = 0$ and is positive at $\xi > 0$. In brief, one needs to learn to solve control problems that involve a variable control accuracy.

4.7. Extension of the Problem: Variable Control Accuracy

We will begin with the examination of the object (4.11). Let it be required to shift the point x to the σ-neighborhood of zero and subsequently keep it there: $\sigma = \sigma(t) > 0$. The prescribed objective will be attained if we ensure at $|x| \geqslant \sigma$ a monotonic decrease of $|x|$ at a rate that weakly exceeds $d\sigma/dt$. Let the functions $\Phi = \Phi(x, t)$ and $\widetilde{\Psi} = \widetilde{\Psi}(|x|, \sigma)$ be defined as shown in Subsecs. 4.2 and 4.3 at all t and let $\Sigma = \Sigma(t) \geqslant |d\sigma/dt|$.

We put

$$u = -k\widetilde{\Psi}(\Phi + \Sigma)\operatorname{sgn} x \tag{4.15}$$

Suppose that a closed system satisfies the conditions of the existence and uniqueness theorem. This requirement imposes definite constraints on the control law, but we will not dwell on this issue here because a more common problem will be met with after a few lectures. It can readily be verified that if $K > B$, the system (4.11), (4.15) moves in the required fashion; the asymptotics of the transient process is studied in the same way as is done in Subsec. 4.4.

The control law of the form (4.15) with the function $\Phi = \Phi(x, \xi, t)$ is also suitable for the system (4.14). To define the function $\sigma = \sigma(\xi, t)$ at $\Sigma = \Sigma(\xi, t)$, we need to use in the law (4.15) the estimates of partial derivatives of σ and the majorant of the phase velocity of the system along the coordinates ξ because

$$\frac{d\sigma}{dt} = \frac{\partial\sigma}{\partial\xi}\dot{\xi} + \frac{\partial\sigma}{\partial t}$$

We will again omit here the analysis for the correctness of equations of the closed system.

4.8. Prospects of Further Extensions

The control problem of the object

$$\dot{x} = \varphi(x, \xi,\, u,\, t) + b(t)u$$

does not contain anything essential if the function $|\varphi|$ admits a uniform upper bound on u, i.e., if we find a function $\Phi = \Phi(x, \xi,\, t)$ such that

$$\Phi(x, \xi,\, t) \geqslant |\varphi(x, \xi,\, u,\, t)|$$

at each value of x, ξ, and t irrespective of u. This means that the effect of the control on the quantity φ is relatively weak. A weaker constraint under which, however, the pattern of deriving the control law remains the same is the following: the majorant $\Phi = \Phi(x, \xi,\, u,\, t)$ of the quantity $|\varphi|$ at all x, ξ, and t displays a lower order of the growth with respect to u than the order of the growth of the linear function, i.e., $\Phi(x, \xi,\, u,\, t) = O(u)$ when $u > 0$.

In all the remaining cases, the problem becomes radically more complex. This problem is unsolvable if we do not introduce additional assumptions for the right side of the object model. The question as to what assumptions are sufficient under the uncertainty conditions is far from trivial; it will not be dealt with in this course.

5. Control of a Nonlinear Multidimensional Object

In this section, we will examine a multidimensional object and look into the possibility of extending the results obtained previously to the technique of control synthesis.

5.1. Model of the Object and Discussion of the Problem

Consider the dynamic system

$$x_i = \varphi_i(x_1, x_2, \ldots, x_n,\, t) + b_i(t)\, u_i; \quad i = 1, 2, \ldots, n \qquad (5.1)$$

In the vector notation, this system (model) obtains the form

$$\dot{x} = \varphi(x,\, t) + B(t)u \qquad (5.2)$$

where $B(t) = \text{diag}[b_1(t), b_2(t), \ldots, b_n(t)]$, x, $u \in R^n$, $\varphi : R^n \times R \to R^n$. For the system (5.2), we state a control problem in which all components of the vector x need be set to tend to zero or made to approach zero. In addition, we think it necessary to ensure a rather fast motion involved in the problem at hand.

One of the ways of deducing the control law is to interpret the n-dimensional object as an aggregate of n one-dimensional objects. Indeed, each equation of the system (5.1) can be treated as a model of a one-dimensional object for control of which the results presented in the preceding section are feasible. Let us recall that for the applicability of these results it suffices: (1) to have the possibility of estimating the majorants of the moduli of all functions φ_i and (2) to know the lower (positive) bound set on the range of changes in the coefficients $b_i(t)$. Repeating the uniform estimations n times, we can obtain a closed system which ensures that a solution falls within the parallelepiped $\{x : |x_i| \leqslant \sigma_i\}$ in a finite time, where σ_i identifies positive constants. However, this procedure of the control synthesis is somewhat cumbersome. Moreover, the procedure becomes still more complex if for the system without a free term in the right side, we have to work out the control that lets $x(t) \to 0$; the difficulties arise from the need for the use of varying values of $\sigma_i = \sigma_i(x_1, \ldots, x_{i-1}, x_{i+1}, \ldots, x_n)$.

To obviate these difficulties, we will rely on the same suppositions as to the object and select another way for the extension of the results presented in the preceding section. We will form a control that affords a monotonic decrease in the norm of the vector of a controllable quantity, which corresponds to a decay of $|x|$ in the one-dimensional case. In the closed system, the norm will play the role of a Lyapunov function.

5.2. Lyapunov Functions

Let us recall that for the closed system $\dot{x} = f(x, t)$ with the steady-state solution $x = 0$, a Lyapunov function is known as a functional $V(x)$ for which $V(x) \geqslant 0$, $V(x) = 0 \Leftrightarrow x = 0$ and $(d/dt)V(x) \leqslant 0$ in view of the system. The following assertion of the Lyapunov lemma is valid: if a Lyapunov function exists, the solution $x = 0$ is stable in the sense of Lyapunov. This result is used to synthesize a control in the following way: if it is necessary to impart the zero solution the property of the stability in the sense of Lyapunov, we preset a certain functional and shape up a control action so as to provide this functional with the properties of the Lyapunov function.

The same technique will also be applied to synthesize a control for the object (5.2): the control law $u = u(x, t)$ should be able to afford a monotonic decrease in the norm of the current value of the solution for the closed system. To substantiate this technique, there is undoubtely no need to refer to the Lyapunov lemma because the decrease in the norm $x(t)$ is, by definition, just the required approach of the point x to zero in the problem at hand. In this case, the Lyapunov lemma in its statement presented above is not applicable at all: it contains only the stability conditions in the sense of Lyapunov, which is evidently not sufficient for us. So, we have recalled Lyapunov functions only with the aim to assign the technique in question to the class of techniques which are known in the theory of automatic control.

5.3. What Norm Must We Select?

As is known, a norm in the linear space L is said to be a functional h such that $h(x) \geqslant 0$, $h(x) = 0 \Leftrightarrow x = 0$, $h(\alpha x) = |\alpha| h(x)$, $h(x + y) \leqslant h(x) + h(y)$, where x and y are elements of the space L and α is a number. Any norm h generates (or induces) a metric in the space L, i. e., prescribes a method for estimating the distance $\rho(x, y)$ between any two points x and y : $\rho(x, y) = h(x - y)$. A value of $h(x)$ is appropriately treated as a length of the radius vector x or as a distance from the point x to the origin of coordinates.

In a finite-dimensional space, for example, in the space R^n of coordinates of the signal x in the system (5.2) under investigation, all norms are equivalent to one another. This implies that for two arbitrary norms h_1 and h_2 there exist positive constants c_1 and c_2 such that $c_1 h_1(x) \leqslant h_2(x) \leqslant c_2 h_1(x)$ for all x. Hence, it follows that if $h(x(t)) \to 0$ holds for a certain norm on the strength of the system of interest, then the same assertion is valid for any other norm. Therefore, during the synthesis of a control, one can use any norm as a Lyapunov function: if there is a way to let this norm tend to zero, then all of the remaining norms will also tend to zero. Note that the property of a monotonic change in $h(x(t))$ does not generally hold after the replacement of a norm.

So, the choice of one norm or another in the space R^n for control purposes only depends on the considerations of convenience. With the fulfillment of the inequality $(d/dt)h(x(t)) < 0$, we will register a decrease in the radius vector $x(t)$, so that "conveniences" in this case must include the possibility of differentiation. In this connection, let us require that a norm be smooth in x. Thus, we leave out of consideration such frequently applicable norms as

$$\|x\|_1 = \max_i |x_i|, \quad \|x\|_2 = \sum_i |x_i|$$

There is an infinite number of smooth norms in R^n, of which we select only the Euclidean norm. To justify our arbitrariness, we point out that the Euclidean norm finds use in everyday human practice from ancient times and reflects the equality of rights of all trends in the real world. Under the adopted assumptions, the trends observed in the state space of the system of interest are also equal in rights.

5.4. Control Synthesis

We will examine the effect of a control action on the character of motion of $\|x(t)\|$. For convenience of the discussion, we introduce into R^n the Euclidean scalar product: $\langle x, y \rangle = \sum_i x_i y_i$. Recall that $\langle x, y \rangle = \|x\| \cdot \|y\| \cdot \cos \alpha$, where α is an angle between the vectors x and y. We have

$$
\begin{aligned}
\frac{d}{dt}\|x\| &= \frac{d}{dt}\sqrt{\langle x, x \rangle} \\
&= \frac{\langle x, \dot{x} \rangle}{\sqrt{\langle x, x \rangle}} \\
&= \left\langle \frac{x}{\|x\|}, \dot{x} \right\rangle \\
&= \left\langle \frac{x}{\|x\|}, \varphi \right\rangle + \left\langle \frac{x}{\|x\|}, \frac{Bu}{\|u\|} \right\rangle \cdot \|u\|
\end{aligned}
\tag{5.3}
$$

Therefore, the quantity $\|x(t)\|$ decreases in magnitude when the direction control vector $u/\|u\|$ ensures the negation, but the norm $\|u\|$ ensures a rather high value of the second summand in expression (5.3). Assume that we know a continuous function $\Phi : R^n \times R \to R$ such that

$$
\Phi(x, t) \geqslant \|\varphi(x, t)\|
\tag{5.4}
$$

at all x and t and also know the constant $b > 0$

$$
\frac{1}{b} \leqslant b_i(t) \leqslant b
\tag{5.5}
$$

at all i and t. Then, it is easy to verify that the control law

$$
u = -\frac{x}{\|x\|} k\Phi(x, t)
\tag{5.6}
$$

secures in the closed system a strict decrease in $\|x\|$ throughout where $\Phi(x, t) > 0$ under the condition $k > b$. Let it be necessary that this condition should hold everywhere at $x = 0$.

But it should be borne in mind that the inferred conclusion of the character of a change in $\|x\|$ has the meaning only if the equations of the closed controllable system are correct. We assume that for this purpose, it is sufficient that the control action (5.6) should be continuous and the local Lipschitz one in x. We first clarify when the control is continuous.

The law (5.6) contains the cofactor $x/\|x\|$ that is discontinuous at zero (this is sgn x in the one-dimensional case). For it to be continuous, the control must go to zero at this point. In turn, for this purpose it is sufficient to fulfill the condition $\Phi(0, t) \equiv 0$. But this identity is compatible with the condition (5.4) only for objects without a free term in the right side of the model.

In the general case, $\Phi(0, t) \not\equiv 0$ and the continuity can be brought about only by introducing a special smoothing cofactor into the control law. We will proceed in the same way as we did in solving the one-dimensional problem. Let $\sigma > 0$ and $\widetilde{\Psi}$ be a continuous function $\widetilde{\Psi} : R \times R \to R$ such that $\widetilde{\Psi}(\|x\|, \sigma) \geqslant 1$ at $\|x\| \geqslant \sigma$ and $\widetilde{\Psi}(0, \sigma) \equiv 0$. Then, the control

$$u = -\frac{x}{\|x\|} \widetilde{\Psi}(\|x\|, \sigma) k \Phi(x, t) \qquad (5.7)$$

is continuous everywhere. If, in addition, the control is the local Lipschitz one in x, then the quantity $x(t)$ actually decreases in the monotonic fashion in the closed system at $\|x\| \geqslant \sigma$.

The question as to when the control (5.7) satisfies the Lipschitz condition in x on any bounded set will be put off for a few lectures until we consider in detail a more common class of objects.

5.5. Character of Transient Processes

According to (5.3), the control (5.7) at $\sigma > 0$ or the law (5.6) at $\sigma = 0$ ensures in the range $\|x\| \geqslant \sigma$ the inequality

$$\frac{d}{dt}\|x\| \leqslant -\left(\frac{k}{b} - 1\right) \Phi(x, t) \qquad (5.8)$$

whence, under the condition

$$\Phi(x, t) \geqslant M\|x\| \qquad (5.9)$$

the following estimate is evident:

$$\|x(t)\| \leqslant \|x(t_0)\| \exp\left\{-\left(\frac{k}{b} - 1\right) M(t - t_0)\right\}$$

It can be assumed that at a fairly high value of k and/or M, this kind of exponential estimate conforms to the requirements of the control problem. The constant $\sigma > 0$ (in the case where $\Phi(0, t) \not\equiv 0$) must be consistent with the desired accuracy of control.

Thus, the constructed control can actually serve as a solution of the stated problem. It should be noted that it is also necessary to provide definite geometric properties of phase trajectories for many control problems. For example, it often so happens that there is a need to preclude overcontrol, i. e., to ensure constant signs of all or some components of the vector x. The problems of this type will be dealt with in the text, but after a few lectures.

5.6. Possibilities for Extension of the Problem

It can readily be seen that we in essence replaced the problem for control of the vector quantity $x \in R^n$ by the problem for control of the scalar $\|x\| \in R$. Therefore, the extensions of the examined problem must correspond to possible extensions of the one-dimensional control problem considered in Sec. 4. This makes it possible to restrict the discussion here only to the enumeration of ways for the transition to more common problems.

First, the dependence of current values of the elements of the matrix B on x, u or any other arguments is of no significance if these elements vary in the interval $[1/b, b]$. In this case, the control law remains invariant.

Second, the dependence of components of φ on additional state parameters ξ makes it necessary to use the majorant $\|\varphi\|$ of the form $\Phi(x, \xi, t)$ in the control law; then, there is no need to change anything.

Third, the desire to solve the control problem with the variable control accuracy σ forces us to strengthen the function Φ in the control law by the majorant of the modulus of the total derivative of σ with respect to time in view of the system.

Fourth, in the case where values of the function φ depend on u, the control problem is solvable if $\|\varphi\|$ increases in $\|u\|$ slower at each point of the state space and at each instant of time than a linear function.

5.7. Comments on the Control Law

The simple formula (5.7) will play a rather important role in our lecture course, for which reason we will briefly outline its features. The control action

$$u = -\frac{x}{\|x\|} k\Phi\widetilde{\Psi}$$

is made up of three cofactors: the direction cofactor $-x/\|x\|$, "force" cofactor $k\Phi$, and smoothing cofactor $\widetilde{\Psi}$, each performing its function.

The direction cofactor $x/\|x\|$ coincides with the direction control vector $u/\|u\|$ at $x \neq 0$; owing to this cofactor, the vector u always points toward the origin of coordinates at each point of the phase space, apart from zero.

The contribution of the control to the right side of the system model

$$\dot{x} = \varphi + Bu$$

namely, the vector Bu already has its direction that changes with time in an unknown manner, but this vector necessarily has the component (projection) directed to the origin of coordinates.

The "force" cofactor $k\Phi$ is a scalar which is so large that at $\|u\| \geqslant k\Phi$ the above-mentioned projection of Bu on the radius vector x exceeds a value of φ. Therefore, the entire right side $\varphi + Bu$ of the system, and not only the contribution of the control, has the component pointing toward zero, but only under the condition $\|u\| \geqslant k\Phi$.

The smoothing cofactor weakens the "force" cofactor near the coordinate origin so as not to allow the point x to "force its way" into its destination spot at a nonzero speed. In other words, this cofactor makes up for the shortcomings of the remaining cofactors at zero: the direction vector here is discontinuous, while the "force" cofactor retains the discontinuity at $\Phi(0, t) \not\equiv 0$.

As a matter of benevolent self-criticism, we can note that our control could have been taken somewhat smaller. First, it would have been well to select the direction vector so that the entire contribution of Bu into the right side, apart from the vector component, should have pointed toward zero. Indeed, why expend control forces in useless directions? Second, the "force" cofactor must "overcome" not the entire quantity φ, but only the projection of this vector on the radius vector x.

However, to satisfy these desires calls for a much greater body of information on φ and B than we had used. Let us agree on the condition that we lack additional data on an object. Therefore, we reject the attempt made to subject the text to the fault-finding and finish this chapter with the feeling of satisfaction.

Control Problem
for an Intermediate Link

In this chapter, we will outline a version of the mathematical implementation of the idea involving the transfer of actions through an intermediate dynamic system. A controllable object will be taken as a pair of the interrelated systems A and B and the control law will be set up so that the system B exerts actions of a special class on the system A. We are to do the following:

(a) substantiate our choice of the class of transferable actions, i. e., explain why we hope to exert a beneficial effect on a controllable quantity as we force the system B to produce an action of the definite type on A;

(b) suggest control algorithms that cause B to operate in the desired fashion;

(c) specify the conditions under which the suggested algorithms do their duty.

Let us note that our concern will lie only in the transfer, as such, of actions, no matter what aftereffects they can induce. It is not all the elements of the above-mentioned class of transmitted actions that can ensure the desired behavior of a controllable quantity. The issues of the search for the required element in this class are treated in Chapter 4.

The transmission of actions represents the most important aspect of the approach developed in this course. Therefore, we have taken it necessary not to make haste in the presentation of the course and supplement it with different appropriate topics.

6. Results — Predecessors

Some control algorithms long known in cybernetics are found to owe their success to the fact that they produce the same effect of transfer of actions as that which is dealt with in this chapter. Although the results obtained in this case can be used only for a relatively narrow class of control problems, nevertheless, they are "ancestors" of the approach set forth here. In this section, we examine features of these results by considering an example of the control problem for a linear second-order object.

6.1. Statement of the Problem: Remote and Immediate Objectives of Control

Consider the object

$$\dot{x}_1 = x_2$$
$$\dot{x}_2 = a_1 x_1 + a_2 x_2 + bu \tag{6.1}$$

Let us assume that the coefficients a_1, a_2, and b are nonmeasurable, vary in the ranges $[-A, A]$ and $[1/B, B]$, respectively, and $A \geqslant 0$ and $B \geqslant 1$ are known numbers. Suppose that we have to work out a control law such that it ensures the motion of $x \rightarrow 0$ in the system (6.1) at $t \rightarrow \infty$ with the required exponential asymptotic behavior (x is a two-dimensional vector with coordinates x_1 and x_2).

The stated character of motion of the closed system specifies the common, "remote" objective of the control synthesis, which we can attain in a variety of ways. We will state a more specific problem, i. e., prescribe an "immediate" objective the attainment of which will afford a means of meeting the requirements of the initial problem for the system. Note that the fulfillment of the condition $x_2 = -cx_1$, where $c > 0$, leads to the condition $\dot{x}_1 = -cx_1$ in view of the first equation of the system (6.1) and causes x to move to zero $x \rightarrow 0$ with the asymptotic behavior of $x = O(e^{-ct})$. At a sufficiently high value of c, this motion can be taken satisfactory. Moreover, along with the required character of solutions of the problem, the above conditions generate two useful effects: first, they preclude overcontrol, i. e., the coordinates x_1 and x_2 do not change sign, and, second, the character of solutions only slightly depends on the parameters a_1, a_2, and b at a fairly low value of Δ, but does not depend at all when $\Delta = 0$.

The above remark can serve as a basis for the selection of an immediate control objective. Let us introduce the designations: $s = x_2 + cx_1$ and $\sigma = \Delta |x_1|$. We will build up a control so that the quantity s can reach the σ-neighborhood of zero in view of the system and remain within it later on.

Thus, the immediate objective is to solve the control problem for the quantity s. In view of (6.1), this variable is described by the differential equation $\dot{s} = a_1 x_1 + a_2 x_2 + bu + cx_2$ or, with allowance made for the definition of s, it is given as

$$\dot{s} = (a_1 - a_2 c - c^2)x_1 + (a_2 + c)s + bu \tag{6.2}$$

We have examined the control problem for an object of the form (6.2) in Sec. 4: this is a one-dimensional control problem with the additional state parameter x_1 and the required variable control accuracy $\sigma = \Delta |x_1|$. By analyzing different approaches to the solution of this problem, we will seek, where possible, correlations between the current issues and the issues considered in the preceding chapter.

6.2. Linear Control

Consider possibilities of the control law of the form

$$u = -ks \tag{6.3}$$

If we recall the definition of s and the first equation of the system (6.1), it will be clear that (6.3) describes what is known as the proportional-differential controller (PD controller (Fig. 6.1)):

$$u = -k\dot{x}_1 - kcx_1$$

We rewrite the law (6.3) in the form $u = -k|s|\,\text{sgn}\,s$. This expression contains the basic features of the control laws derived in the preceding chapter: it includes the direction cofactor $\text{sgn}\,s$ and the "force" cofactor $k|s|$. The latter cofactor goes to zero at the discontinuity points of $\text{sgn}\,s$, and so this expression does not contain a special smoothing cofactor.

We will investigate the issue of when $k|s|$ actually takes up its "force" duties. When $s = 0$, we have

$$\frac{d}{dt}|s| = (a_1 - c(a_2 + c))x_1\,\text{sgn}\,s + (a_2 + c)|s| - bk|s| \tag{6.4}$$

In the right side of (6.4), the summand of order x_1 is majorized by the quantity of order s only where $x_1 = O(s)$; such a relation exists at $|s| \geqslant \sigma = \Delta|x_1|$ if $\Delta > 0$. In this case, $|x_1| \leqslant (1/\Delta)|s|$ and the control (6.3) makes the quantity s decrease steadily if the coefficient k is fairly high. In view of the system, we have

$$|\dot{\sigma}| = \Delta|x_2| \leqslant \Delta|s| + c\Delta|x_1| \leqslant (\Delta + c)|s|$$

Therefore, at a rather high value of k, the control ensures that at $|s| \geqslant \sigma$ the quantity $|s|$ diminishes at a rate that exceeds the rate of a change in the current value of σ. As a result, the system either reaches the domain $G = \{x : |s| \leqslant \sigma\}$ in a finite time and remains there or tends to zero beyond G.

From the above line of reasoning, we can obtain the sufficient conditions for the desired motion:

$$bk > |a_1 - c(a_2 + c)| \cdot \frac{1}{\Delta} + |a_2 + c| + (\Delta + c).$$

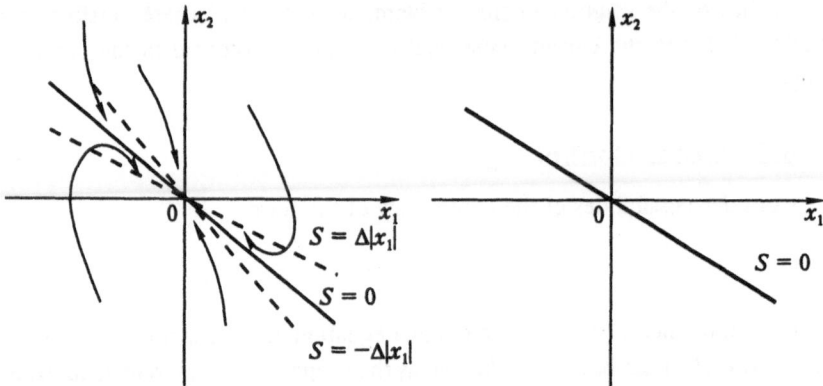

Fig. 6.1. The proportional-differential con- trol law (PD controller) $u = -ks$ generates the phase portraits displayed in the figure, where $k = \text{const} > 0$ and $s = x_2 + cx_1$. Under almost all initial conditions, the point x moves in the domain $|s| \leqslant \Delta|x_1|$ in a finite time, where the motion at low values of Δ depends little on indefinite parameters of the model. However, $\Delta \to 0$ only when $k \to \infty$.

Fig. 6.2. Strong feedback obtained through the passage to the limit of $k \to \infty$ in the PD controller forces the point x to shift instantly to the straight line $s = 0$ and then move to zero along this line. This motion does not in any way depend on the model parameters. In practice, strong feedback is unrealizable: here, $u = \infty$ at $s \neq 0$.

Therefore, the sufficient estimate of k has the form

$$k > B \frac{(A + C)C + A}{\Delta + A + \Delta + 2c}$$

Note that for this inequality to be valid as Δ decreases, it is necessary to increase k. A simple additional analysis (which is left out) reveals that the above condition is necessary and $k \to \infty$ when $\Delta \to 0$. Hence, we have to increase k and thus to pay for the weakening of the effect of the parameters a_1, a_2, and b on the properties of the transient (for which purpose, we reduce Δ).

6.3. Strong Feedback

The passage to the limit of $k \to \infty$ for the control law (6.3) is known to entail strong feedback. In general, the passage of this type, its correctness, and its properties are treated in the context of the theory of differential equations with low-level parameters for derivatives. The essence of this theory will not be dealt with here. As regards the system (6.1), (6.3), this type of low-level parameter appears in an equation that specifies the motion of the coordinate x_2:

$$\frac{1}{k}\dot{x}_2 = \frac{a_1 x_1 + a_2 x_2}{k} - bs$$

In the limit $k \to \infty$, this equation degenerates into the condition $s = 0$ which, as shown above, induces an exponential decrease in x_1 and x_2.

A strong feedback system behaves in the following manner: a representative point on the plane x_1, x_2 instantly shifts to the straight line $s = 0$ under any initial conditions, as it moves along the axis x_2, and then remains on this line (Fig. 6.2). One cannot but acknowledge that this is quite a significant process. Note that the properties of solutions of the closed system are totally independent of the parameters a_1, a_2, and b, i.e., the disturbances applied to the system.

However, the price payed for this transient process is excessively high. The matter is not even that we need to extend the theory of ordinary differential equations so as to substantiate the passage of $k \to \infty$; this is just done in the theory of equations with low-level parameters. The state of affairs lies in practice. Indeed, according to (6.3), $|u| \to \infty$ when $k \to \infty$ and $s \neq 0$. For this reason, we have to apply an infinite control action to the object the initial state of which does not conform to the condition $s = 0$. This is certainly cannot be done in a real system.

Hence, it will be necessary to give up strong feedback. There are two ways of the digression: first, we can content ourselves with a linear control involving the constraint k and, second, investigate the passage of $k \to \infty$ if the constraints on a value of the control action are available. The first of these ways is treated in the preceding subsection. We now turn to the second way.

6.4. Bounded Actions

Let us first point out that we describe in this text the approach to the control synthesis under the assumptions that the bounds on the coordinates of a state and the coordinates of a control are absent. In fact, these bounds always exist. The approach developed here can be brought in agreement with reality if we assume that these bounds lie rather far away from each other and there is no need to express concern over them, the more so as the nonlinear indefinite problems pose certain questions regardless of these bounds. So, we have begun to speak of the bounds on a control only because we consider extraneous approaches.

In general, a bounded control is efficient only within the bounded domain of changes in the coordinates of a system. The further discussion will be held under the assumption that the system of interest lies within this domain. Let a constraint on the control under study has the form $|u| \leqslant u_{max}$ (Fig. 6.3). Then, we cannot elaborate the control law (6.3) in the case where $|s| \geqslant u_{max}/k$. Instead of (6.3), the following law will in fact work:

$$u = -\min\{k|s|, u_{max}\} \operatorname{sgn} s \qquad (6.5)$$

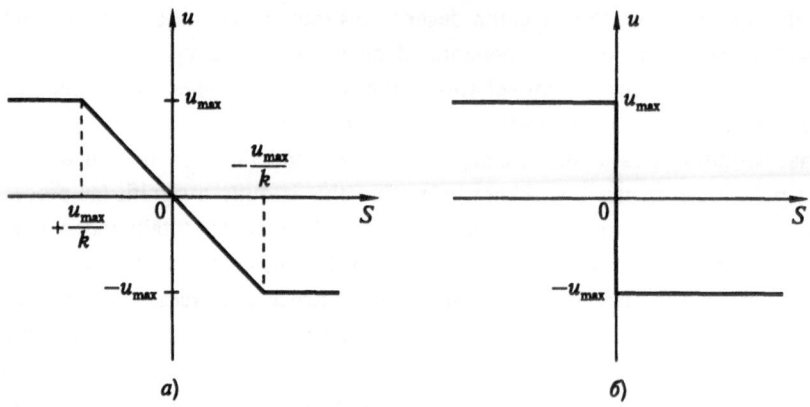

Fig. 6.3. In practice, controls are always bounded. If $|u| \leqslant u_{max}$, then the law illustrated in Fig. 6.3a actually works rather than a PD controller. It is possible to disregard the bounds placed on u only if the domain of the state space under consideration is defined by values of $|s| < u_{max}/k$. However, $u_{max}/k \to 0$ when $k \to \infty$, and so instead of strong feedback, the control $u = -u_{max}\,\mathrm{sgn}\,s$ results as one passes to the limit (Fig. 6.3b). This law does not ensure the instantaneous transfer of the point x to the straight line $s = 0$. It is impossible to implement such a system. In fact, the point x oscillates about the straight line $s = 0$ and the control changes over from a value of u_{max} to $-u_{max}$, and vice versa, at a rate that complies with the limit of system hardware capabilities.

Passing to the limit for $k \to \infty$, we obtain

$$u = -u_{max}\,\mathrm{sgn}\,s \qquad (6.6)$$

The study of the closed system (6.1), (6.6) can be made only on condition that it is possible to redefine correctly the system model by imparting a certain sense to the ordinary differential equations with discontinuous right sides. This theoretical difficulty is overcome in mathematics. It is the severe difficulties of practical character that remain in force: in a real system, a control action will undergo high-frequency oscillations at an amplitude of u_{max}. These oscillations give rise to rapid wear of the actuators of the controllable system, and the identity $s = 0$ does not hold all the same.

There are two points of view on discontinuous controls. First, these controls can in principle be discarded. This position will ultimately be taken up in the approach under development. Second, it is possible to accept the discontinuity, as such, and place emphasis on the elimination of some defects of the law (6.6). We will hold the latter viewpoint in regard to the theory of varying structure systems (VSS) and examine them from this viewpoint in the subsection that follows.

6.5. Varying Structure Systems

In the framework of the theory of systems with varying structures, a control law for the object (6.1) assumes the form

$$u = -k|x_1|\operatorname{sgn} s \qquad (6.7)$$

If the absolute value of the action (6.3) has majorized everywhere the summand of order s in expression (6.2) for the rate of changes in the signal s, then the law (6.7) uses the majorant of another summand of order x_1. The modulus of the law (6.7) with a rather high value of k will majorize both of the summands if the system at hand lies in a domain where $s = O(x_1)$. Therefore, in the course of the analysis, we have to make sure that the system enters this domain from any initial point. Thereafter, the representative point of the system will begin to approach the straight line $s = 0$. Near this

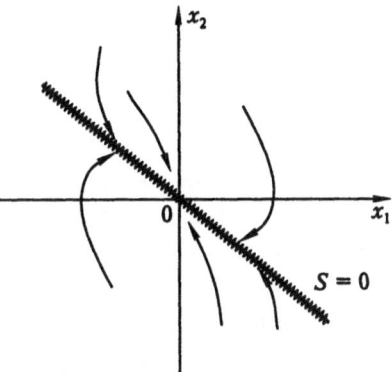

Fig. 6.4. The varying structure system shifts the point x to the straight line $s = 0$ under almost all initial conditions in a finite time. The point then moves in the sliding mode (the "saw" in the figure identifies this way of motion). In practice a control signal undergoes oscillations at an amplitude of $\sim |x_1|$ and at the highest possible frequency.

line, a value of $|s|$ decreases in the varying structure system (VSS), which is not the case for a system with the PD controller, for which reason the condition $s = 0$, once it emerges in the system, will no longer break down at any time (Fig. 6.4).

We will perform the analysis of a closed system. When $s > 0$ and $x_2 > 0$, we get

$$\dot{x}_2 - A\dot{x}_1 = a_1 x_1 + (a_2 - A)x_2 - bk|x_1| \leqslant (A - bk)|x_1|$$

and the quantity $x_2 - Ax_1$ steadily decreases in magnitude if $k > AB$. Therefore, the system either goes out of the domain $s > 0$, $x_2 > Ax_1$ and falls on the straight line $s = 0$ or moves in the domain $x_2 \leqslant Ax_1$, or tends asymptotically to zero as it stays within the domain $s > 0$, $x_1 < 0$. In a similar way, we can also examine the motion of the system in the domain $s < 0$, $x_2 < 0$.

Thus, when $k > BA$, the representative point of the system, which lies on the plane x_1, x_2, either falls within the domain $|s| \leqslant (c + A)|x_1|$ or asymptotically tends to zero. In the latter case, the point obviously moves faster that it does along the straight line $s = 0$, and so we will consider this outcome to be satisfactory. Let us look at the first case: the system goes in the domain $|s| \leqslant (c + A)|x_1|$.

Here,

$$\frac{d}{dt}|s| \leqslant (A + c(A + c))|x_1| + (A + c)^2|x_1| - bk|x_1|$$

Hence, if

$$k > B(A + (A + c)(A + 2c)) \qquad (6.8)$$

the system approaches the straight line $s = 0$. The final result of our analysis is as follows: under the condition (6.8), the system (6.1), (6.7) either reaches the straight line $s = 0$ in a finite time and moves along it to zero (in which case, the system is said to operate in the sliding mode) or tends to zero beyond this line.

The most important merit of the control law (6.7) as against (6.6) is that $u \to 0$ tends to zero as the system approaches the coordinate origin. Therefore, a low control error (i. e., the closeness of x_1 to zero) generates a small control action in the VSS, whereas the law (6.6) shapes up the maximum possible response to any deviation from the straight line $s = 0$.

Nevertheless, the execution of the sliding mode involves considerable practical difficulties, and so it is desirable to derive an acceptable continuous control law. This purpose is achieved in the context of the theory of binary systems.

6.6. Dynamic Binary Control

The following algorithm describes a continuous control:

$$u = -\varkappa|x_1| \quad \begin{cases} \alpha \, \text{sgn} \, s & \text{at } |\varkappa| \leqslant k \\ 0 & \text{at } |\varkappa| > k \end{cases} \qquad (6.9)$$

where α and k are positive constants and $\varkappa$ identifies the quantity that reaches its boundary value and remains invariable until s changes sign.

We call this algorithm the binary and dynamic one, thus pointing to its design features which do not relate to the topic discussed in this section. However, we will make a small terminological digression from the topic. The term "binary" points to the following interpretation of the definition of (6.9): the conversion of $x \mapsto u$ is effected by the operator $u_\varkappa(x)$ with the parameter $\varkappa$ (according to the first equality of (6.9)). The current value of the parameter $\varkappa$ depends on x (according to the second equality of (6.9)). The binary aspects in general and the binary aspects of the algorithms presented in the text will be given more detailed treatment in Sec. 15. The term "dynamic" indicates that the conversion of $x \mapsto \varkappa$ corresponds, in view of (6.9), to a certain dynamic system rather than to a function. In this respect, (6.9) does not belong to the class of control laws among which we have settled to seek means for the "world improvement" in Sec. 3.

We now turn to the investigation of the properties of the control (6.9) (Fig. 6.5). Let the initial value of $\varkappa$ fall within the interval $[-k, k]$. Then, at any initial values of x_1 and x_2, either the condition $\varkappa = k \operatorname{sgn} s$ or the condition $s = 0$ will appear in the system (6.1), (6.9) in a time that does not exceed $2k/\alpha$. Obviously, the first of these conditions converts (6.9) to (6.7) and ensures that x tends to zero or s goes to zero in a finite time when k is sufficiently high (for example, when (6.8) is valid).

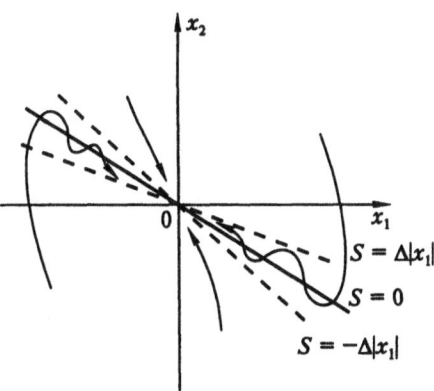

Fig. 6.5. The dynamic binary control transfers the point x to the domain $|s| \leqslant \Delta|x_1|$ under almost all initial conditions in a finite time. In this case, the control action oscillates at an amplitude of $\sim |x_1|$ and a limited frequency specified by the control algorithm parameters. In contrast to the PD control law, the quantity $|u|/|x_1|$ is limited by a constant that does not grow as Δ decreases

Let us look at a motion that begins from the straight line $s = 0$. As it leaves the line, the system will have an opportunity either to return to this line or to convert to the VSS in the time $T \leqslant 2k/\alpha$. We will verify the conditions under which these motions occur within the bounds of the domain $G = \{x : |s| \leqslant \Delta|x_1|\}$ and outside the straight line $s = 0$. Here, it is evidently sufficient to require that, first, the system should not be able to leave this domain in the time $2k/\alpha$ and, second, the control law (6.7) should impart the given domain the invariance property. Recall that the invariant domains are known to be those domains which the system is unable to leave.

In view of (6.9), we have $|bu| \leqslant Bk|x_1|$ and so at $x_1 \neq 0$ we obtain $x_1 \neq 0$

$$\frac{d}{dt} \frac{|s|}{|x_1|} \leqslant A + Bk + (A + c + \Delta)(c + \Delta)$$

Consequently, if

$$\alpha > \frac{2k}{\Delta}[A + Bk + (A + c + \Delta)(c + \Delta)] \tag{6.10}$$

the control (6.9) converts to (6.7) before the system leaves the domain G. This domain is invariant if $|s| - \Delta|x_1|$ diminishes at $|s| = \Delta|x_1|$. At $|s| \geqslant \Delta|x_1|$, we have

$$\frac{d}{dt}|s| - \frac{d}{dt}\Delta|x_1| \leqslant \left(A + (A + c + \Delta)(c + \Delta) - \frac{k}{B}\right)|x_1|$$

Therefore, for G to be invariant, it is sufficient to fulfill the condition

$$k > B[A + (A + c + \Delta)(c + \Delta)].$$

In view of (6.8), we find that if the inequality

$$k > B[A + (A + c + \Delta)(A + 2c)]$$

and the inequality (6.10) are valid, the system (6.1), (6.9) either reaches the domain G in a finite time and goes to zero within it or tends to zero beyond G.

An advantage of the binary control over the PD control law is that there is no need for the coefficient k in (6.9) to grow infinitely as Δ decreases; it is only the rate of a change in Δ that increases here.

7. New Properties of Two-Dimensional Problems

We will put off the discussion of the general aspects of the approach under development until the next section, but now continue the treatment of two-dimensional problems. We will show how our approach works to attain the same "immediate" control objective (and hence the "remote" objective) as that attained by its predecessors considered in Sec. 6, but for a wider class of objects.

7.1. Discussion of Statement of the Problem

Consider the object model

$$\begin{aligned}
\dot{x}_1 &= x_2 \\
\dot{x}_2 &= \varphi(x, t) + bu
\end{aligned} \tag{7.1}$$

a specific case of which is the model (6.1). We will infer again that the coefficients $b = b(t)$ vary in the interval $[1/B, B]$, where $B > 1$. The two-dimensional vector x will be thought of as a control error signal. Hence, as before, the remote control objective is to ensure that $x \to 0$ with the desired asymptotic behavior of the transient.

We will also retain the immediate control objective, considering that the validity of the inequality

$$|s| \leqslant \sigma \tag{7.2}$$

offers the sufficient condition under which the required character of motions appears in the system, where $s = x_2 + cx_1, \sigma = \Delta|x_1|$, and $c > \Delta > 0$. As in the system (6.1), this circumstance depends on the form of the first equation for the object, which describes the dynamics of the coordinate x_1. The second equation in (7.1) specifies the presence (or absence) of the possibilities of fulfilling (7.2) and also the means necessary to retain the above inequality.

It is easy to verify that none of the algorithms examined in the preceding section is able to ensure inequality (7.2) if the object (7.1) does not degenerate into (6.1). Indeed, all the control laws considered before satisfied the condition $u = O(\|x\|)$, for which reason the action u could force the coordinate x_2 (and hence s) to move in the required direction only if $\varphi(x, t) = O(\|x\|)$. The functions φ conforming to this requirement can be given in the linear form $a_1 x_1 + a_2 x_2$ with limited coefficients; here, (7.1) coincides with (6.1). In the general case, we need to synthesize a new control law for the object (7.1).

7.2. Control Synthesis

In view of the system of interest, the quantity s varies according to the equation

$$\dot{s} = c x_2 + \varphi(x, t) + b u \tag{7.3}$$

The problem at hand is to stabilize the one-dimensional system (7.3) at zero with the control accuracy σ in the presence of the additional parameter x that has an impact both on the dynamics of the controllable quantity s and on a value of the varying accuracy σ. The problems of this type were considered in Sec. 4. So that the subjects under study will fully comply with the subjects discussed in Sec. 4, it would be desirable to pass from the variables x_1 and x_2 in the right side of (7.3) to x_1 and s by replacing $s = x_2 + c x_1$. We will not perform this replacement because it is in essence useless.

According to the results presented in Sec. 4, the control law for (7.3) can have the form

$$u = -k \, \widetilde{\Psi}(|s|, \sigma) \, F(x, t) \, \text{sgn} \, s, \tag{7.4}$$

where $F(x, t) \geqslant |\varphi(x, t) + c x_2| + \Delta |x_2|$ at $k > B$, $\widetilde{\Psi}(0, \sigma) \equiv 0$, $\widetilde{\Psi}(|s|, \sigma) \geqslant 1$.
For $|s| \geqslant \sigma$ we can use, for example, the functions

$$\min\left\{1, \frac{|s|}{\sigma}\right\}, \quad \frac{2|s|}{|s| + \sigma}, \quad \sqrt{2}|s| \frac{1}{s^2 + \sigma^2}$$

Let us define the majorant of the components in the right side of the object model:

$$\Phi(x, t) \geqslant \max\{|x_2|, |\varphi(x, t)|\}$$

In this case, the control law can be given as

$$u = -k \, (1 + \Delta + c) \, \Phi(x, t) \, \widetilde{\Psi}(|s|, \sigma) \, \text{sgn} \, s \tag{7.5}$$

7.3. Correctness Problem

The control law should be set up so that equations for a closed system satisfy the conditions of the existence and uniquency theorem for the solutions of ordinary differential equations. Assume that equations (7.1) satisfy these conditions at $u = $ const. Then, the control law (7.5) must prescribe the function $u = u(x,t)$ which is the continuous one in x and t and the local Lipschitz one in x. In this case, the solution of (7.1), (7.5) is existent and unique for any initial point, and so it makes sense to carry out the investigation, namely, the qualitative analysis of the properties of these solutions.

Let us clarify when the function $u(x,t)$ of (7.5) displays these properties. We will not check the Lipschitz condition specifically for this function because it will be studied later on in a more general way. We will dwell on the analysis of the control continuity which, as noted above, is not only the condition of the formal correctness of equations, but is also related to the possibilities for the practical implementation of the system.

The cofactor $\widetilde{\Psi}$, used in (7.5) to smooth out the discontinuities of the function sgn s, itself undergoes the discontinuity at $s = \sigma = 0$. In the case under study, this condition holds at the point $x = 0$. Therefore, even with the use everywhere of the continuous function $\Phi(x,t)$, the control (7.5) proves discontinuous at the coordinate origin whenever the condition $\Phi(0,t)$ breaks down. But for this condition to be valid, it is necessary to provide the identity $\varphi(0,t) \equiv 0$. In the general case, the absence of this identity makes it necessary to consider the problem for ensuring the control continuity at zero.

7.4. Control in the General Case

In developing the control algorithm in Sec. 4 for the case where $\varphi(0,t) \not\equiv 0$, we used the following approach: the control accuracy was preset so that it always took on a positive value. As regards the control problem for the object (7.3), this accuracy corresponds to a value of the function $\sigma = \sigma(x_1)$. We set it positive everywhere. Let, for example,

$$\sigma = \max\{\delta,\ \Delta|x_1|\} \tag{7.6}$$

Here, we assume that $\delta = 0$ if $\Phi(0,t) \equiv 0$ (thus obtaining the function $\sigma(x_1)$ used before) and, otherwise $\delta > 0$. All the remaining elements of the control law described above can be left intact.

7.5. Behavior of Solutions

It can readily be seen that if the closed system of differential equations is correct, then the quantity s decreases at $|s| \geqslant \sigma$, which does so at a higher rate than the rate of changes in σ. Therefore, the domain $G = \{x : |s| \leqslant \sigma\}$ in the system (7.1), (7.5) is invariant. A decrease in s can be viewed as the approach of the system to G. Strictly speaking, in the adopted assumptions we have to do with the "stationarity" rather than with the approach of the system. Indeed, the system can come to a halt beyond the domain G if there are zeros of the function $\Phi(x, t)$ beyond G. It is possible to introduce an additional condition $\Phi(x, t) > 0$ at $x \neq 0$, but this condition is insufficient as it is in the one-dimensional case (see Sec. 4). In addition, we should take it that values of Φ at each point cannot decrease rapidly.

Assume that $\Phi(x, t)/\|x\| \geqslant \mathrm{const} > 0$. In this case, outside G and on ∂G we have

$$\frac{d}{dt}|s| + \left|\frac{d}{dt}\sigma\right| \leqslant (\varphi(x, t) + cx_2)\,\mathrm{sgn}\,s + \Delta|x_2| - bk(1 + c + \Delta)\,\Phi(x, t)$$

$$\leqslant -\left(\frac{k}{B} - 1\right)(1 + c + \Delta)\,\Phi(x, t)$$

Therefore, the system approaches G and either gains access to this domain in a finite time or always remains outside it. The latter event can occur only if $x \to 0$ beyond G; this type of motion arises under certain initial conditions of the systems, such that $\Phi(0, t) \equiv 0$ in (7.5) and $\delta = 0$ in (7.6).

Let us examine in more detail the motion along the trajectories entirely extending beyond G, which will be called exceptional trajecties. It is easy to check that the quantity $|x_2| - (c + \Delta)|x_1|$ does not increase monotonically in the closed system beyond G. Hence, the exceptional trajectories can lie only within the domain $|x_2| \geqslant (c + \Delta)|x_1|$. If a certain trajectory totally lies in this domain, then it is obviously an exceptional one. In view of the first equation of (7.1), the motion along this trajectory obeys the condition

$$|x_1| = O(\exp(-(c + \Delta)t))$$

At a fairly high value of c, we will consider this motion to be satisfactory.

Let us now look at all other trajectories which are not exceptional ones. The system penetrates in the domain G along them in a finite time, following which it never leaves the domain. Here, if $\delta = 0$ in (7.6), then $\|x\| = O(\exp(-(c - \Delta)t))$.

We will agree to consider this asymptotic behavior satisfactory. If $\delta > 0$, the system moves in the same way as it does at $\delta = 0$ within the domain G beyond the parallelogram $|s| \leqslant \delta, |x_1| \leqslant \delta/\Delta$. Following the inevitable penetration of the point x into this parallelogram, the character of motion changes and, possibly, there is no way to ensure that $x \to 0$. One has to put up with this fact since the sacrifice here is not too large: the above parallelogram is an invariant set of the closed system and the appropriate selection of δ permits us to place it into the preassigned neighborhood of the coordinate origin, no matter how small.

Thus, after the requisite selection of adjustable parameters, the control law (7.5) ensures either the motion in the domain G as some time elapses, which tends to zero or to the prescribed neighborhood of zero with the desired exponential asymptotics, or provides the motion of $x \to 0$ outside G.

7.6. Some Remarks

The considered example of the second order enables us to display the basic features of the approach under study to the control synthesis for objects of an appreciably wider class. We will note the most important features of the approach.

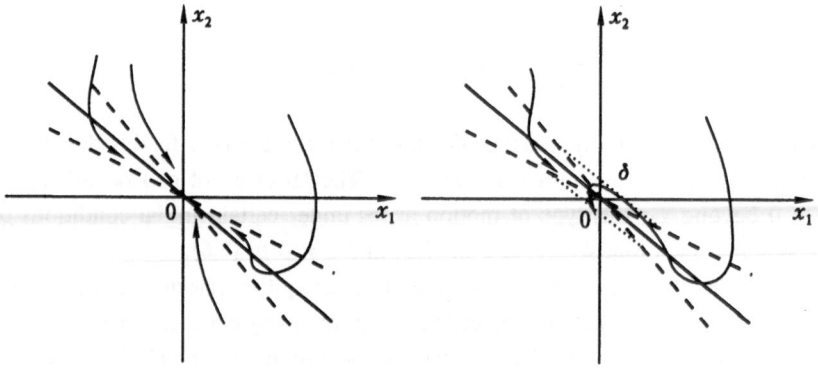

Fig. 7.1. The suggested control law, just like the PD controller and the dynamic binary algorithm, forces the system without the free term in its right side to move in the domain $|s| \leqslant \Delta|x_1|$ after a certain instant of time under all initial conditions. Here, a decrease in Δ does not call for an increase in $|u|$ and the control action does not cause oscillations of s. But the principal aspect is that the suggested law is suitable for nonlinear unstable indefinite objects.

Fig. 7.2. The suggested control law with free term ensures the fulfillment of the condition $|s| \leqslant \max\{\delta, \Delta|x_1|\}$, after some time, where $\delta > 0$. Here, the control does not cause $x \to 0$, but the proper selection of δ makes it possible to limit the system motion within as small the neighborhood of the coordinate origin as one likes.

1. A control action aims to bind the system coordinates together through a definite relation that takes the form of a functional relation resolved approximately (in some strict sense of the word). In principle, the accuracy of fulfilling this functional relation depends on the possibilities of the continuous control available in the system. Thus, if $\Phi(0, t) \equiv 0$, it can be taken that $\delta = 0$ in (7.6); otherwise, one has to put $\delta > 0$.

2. Among all the relations of the above type, the choice is made of a relation that serves to afford the sufficient condition for the emergence of desired motions in the system, regardless of what means (controls) are chosen to effect these motions.

3. The initial problem is replaced by two problems. The objective of the first problem is to search a functional relation that provides the desired character of motions. The second serves to work out the control that puts into effect the found relation.

In Fig. 7.1 is shown the diagram that illustrates the suggested control law; the diagram illustrating this law with a free term is shown in Fig. 7.2.

8. Internal Feedback

At the beginning of the lecture course, we dealt with the interrelation between different systems and pointed to the possibility of using this interrelation so as to "improve the world". In the text given below, we specify this line of reasoning. The "world improvement" is taken to mean a solution of the automatic control problem and all the interacting systems are given in terms of ordinary differential equations. We will show how to "cut" a controllable object into interrelated dynamic systems and what actions transferred through these systems have an actual chance to stimulate the desired behavior of a controllable quantity.

8.1. Natural Control Loops

Consider the closed (input-free) dynamic vector system

$$\dot{x} = f(x, t) \tag{8.1}$$

where $x \in R^n$ and $f : R^n \times R \rightarrow R^n$. We arbitrarily break up the set of coordinates of the vector x into two subsets, put in order their elements (also arbitrarily), and form two vectors x_A and x_B from them. Hence, we decompose the vector x into a sum of two orthogonal projections which, in view of (8.1), obey the equations

$$\begin{aligned} \dot{x}_A &= f_A(x_A, x_B, t) \\ \dot{x}_B &= f_B(x_A, x_B, t) \end{aligned} \tag{8.2}$$

Equations (8.2) and (8.1) are obviously given in different vector notations of one and the same system of n scalar equations:

$$\dot{x}_i = f_i(x_1, x_2, \ldots, x_n, t); \quad i = 1, 2, \ldots, n$$

The notation (8.2) corresponds to the representation of the system (8.1) as a pair of interrelated systems A and B. The system A converts the input x_B to the output x_A (the first equation of (8.2)) and the system B converts its input x_A to the output x_B (the second equation of (8.2)). The properties of the solution $x(t)$ of (8.1) or the properties of its projections $x_A(t)$ and $x_B(t)$ can be thought of as the result of the interaction between the systems A and B. For now, this interaction appears symmetric: the systems A and B are formally equivalent. Let us disturb this symmetry, namely, look at B "from the viewpoint" of the system A. The input action for the system A (the vector x_B) is the output of the system B and represents the result of the conversion by the system B of the output x_A of the system A. In brief, the system B converts the output of the system A to the input of the latter system. In the language of the theory of automatic control, it can be said that the system B specifies a feedback loop built around the dynamic system A. We indicate that this is an internal feedback loop of the dynamic system (8.1) because it converts a portion of the coordinates of the system to other of its coordinates.

The introduced concept allows us to interpret an arbitrary vector system as a closed control loop: the feedback loop defined by the second equation of (8.2) is applied around the dynamic system defined by the first equation of (8.2). Who or what has created this loop? It owes its origin to nature, history, or the designer if (8.2) is the model of a natural, a social, or a technical system, respectively. The properties of the solutions $x_A(t)$ and $x_B(t)$ of equations (8.2) are taken as being the result of the application of feedback to the system A.

8.2. Internal Feedbacks and Control Problems

If treatment is given to a ready-made closed system, the introduced notion of internal feedback only serves to interpret the behavior of this system. It is quite possible that there is no small sense at all in this interpretation. However, in our lecture course we examine the problem of the "world improvement", i.e., the control problem. Therefore, it is not sufficient for us to have a purely research result. We will reveal how the concept of internal feedback applies to control problems.

Suppose there exists a controllable vector object

$$\dot{x} = g(x, u, t) \qquad (8.3)$$

From the formal viewpoint, the problem under investigation is to specify the function $u = u(x, t)$ which, on substituting it into the right side of (8.3), would convert (8.3) to the vector equation whose solution complies with certain adopted requirements. It can be said that (8.3) is a "semiproduct" of the system of the form (8.1). The control $u(x, t)$ fills out the "vacant site" in the right side of (8.3) and finishes off (8.3) so that it becomes the system with the requisite properties. In a similar way, a satellite (or a nuclear warhead) supplements a missile-carrier to produce the desired functional object.

We decompose the vector x into a sum of two projections and represent the model as before by a system of the equations (Fig. 8.1)

$$\dot{x}_A = g_A(x_A, x_B, u, t)$$
$$\dot{x}_B = g_B(x_A, x_B, u, t)$$

(8.4)

It will be said that the first equation of (8.4) corresponds to the dynamic system A, i.e., the subsystem of the initial object, and the second equation corresponds to the system B, i.e., the internal feedback loop placed around A. In this interpretation, the function $u(x, t)$ finishes building both the system A and the

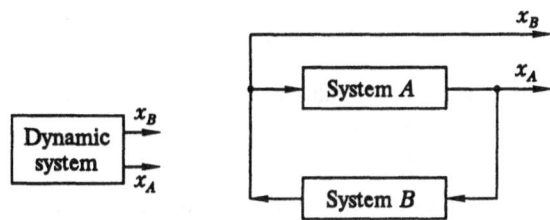

Fig. 8.1. The dynamic system in the form of a pair of interrelated systems A and B admits the following interpretation: the system A has the feedback loop defined by the system B (the internal feedback loop of the initial system).

internal feedback operator. In the case of the favorable outcome of "building", this feedback loop exerts the desired action on the system A.

It is necessary to clarify the following. We assume that the initial control problem involves requirements imposed on the vector function, such that we can represent them in the form of the conjunction of the requirements on $x_A(t)$ and $x_B(t)$. Obviously, this condition limits the class of control problems under consideration. For example, to this class belongs the problem in which the requirement is that the quantity $\sum_i |x_i(t)|$ rather than $\sum_i x_i(t)$ must tend to zero. The tracking problem, which is of prime interest to us, belongs to this class (recall that in the tracking problem, the controllable quantity must reproduce current values of a certain assigned vector variable).

What is the way to finish off the systems A and B? To answer this equation, we can first examine various versions of the "construction" result, then select such A and B that they fit each other, i.e., jointly provide the desired character of motions of $x(t)$, and, finally, determine the function $u(x,t)$ that implements the selected version.

In this case, a control can play one of the three roles. First, the function $u(x,t)$ can place the system A in correspondence with the assigned internal feedback loop, i.e., convert A to such a system that the system B produces the desired effect on A. In this case, the following model is valid:

$$\begin{aligned} \dot{x}_A &= g_A(x_A, x_B, u, t) \\ \dot{x}_B &= g_B(x_A, x_B, t) \end{aligned} \qquad (8.5)$$

Second, the control can put the internal feedback operator in correspondence with the prescribed system A:

$$\begin{aligned} \dot{x}_A &= g_A(x_A, x_B, t) \\ \dot{x}_B &= g_B(x_A, x_B, u, t) \end{aligned} \qquad (8.6)$$

Third, the function $u(x,t)$ can simultaneously convert A and B to systems such that they conform to each other in the above sense. Here, the equations assume the most common form (8.4).

We consider in more detail these three versions of operation of the control. The first two versions are totally symmetric to each other and one converts to the other on the replacement of designations. Therefore, one of the versions that corresponds to (8.5) will be left out of consideration without the loss of generality. In the third version, we have to limit this generality. In the latter case, one and the same function $u(x,t)$ must afford the simultaneous solution of two different problems, which, however, can be beyond the scope of its capability. We will not analyze all possibilities and hence introduce a simplifying assumption. Assume that it is possible to decompose the vector u into a sum of two orthogonal projections u_A and u_B so that (8.4) obtains the form

$$\begin{aligned} \dot{x}_A &= g_A(x_A, x_B, u_A, t) \\ \dot{x}_B &= g_B(x_A, x_B, u_A, u_B, t) \end{aligned} \qquad (8.7)$$

Here, the projection u_A can be used to "finish off" the system A and the projection u_B can serve to complete the operator so as to place the feedback loop around A. The system (8.6) is obviously a particular case of (8.7).

To embark on the implementation of the functions $u_A(x,t)$ and $u_B(x,t)$, we begin with a simple specific case where the projection u_A is absent or where the function $u_A(x,t)$ is found and substituted into the right sides of the equations. A controllable object can be defined by equations of the form (8.6).

8.3. Induced Internal Feedback

Let us look at (8.6) (Fig. 8.2). In the case under study, the control u must convert the system B (the second equation of (8.6)) to a certain internal feedback loop that imparts the desired properties to the system A (the first equation of (8.6)) and shapes along its trajectories $x_A(t)$ a vector function $x_B(t)$ which also displays the required properties. We first raise the question of what form it is necessary to impart to A through its conversion and then clarify how to do this.

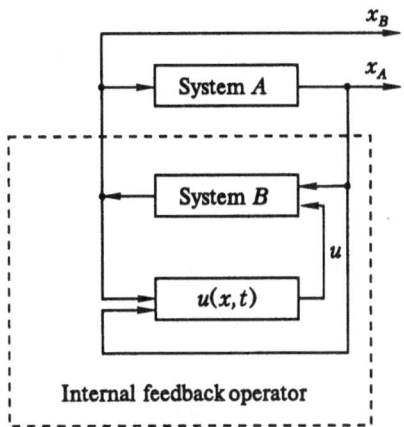

Fig. 8.2. Operation of an internal feedback loop can depend on the control $u = u(x,t)$. The function $u(x,t)$ is thought to be found in such a way that the internal feedback operator on the whole proves close to the prescribed function $v(x_A,t)$, i.e. the control can induce internal feedback with the operator v at a certain error.

The desired internal feedback loop of the object (8.6) is the most ordinary "external" feedback loop for the system A, which involves the conversion of the output x_A of this system to its input x_B. The feedback loop $x_A \mapsto x_B$ must secure a definite character of operation of the system A. To determine this feedback loop, we have to solve the automatic control problem for the system A. In this problem, the controllable object satisfies the equation

$$\dot{x}_A = g_A(x_A, x_B, t)$$

where the vector x_B serves as a control. This problem differs from most of the automatic control problems in that it includes an additional requirement: the control action x_B must vary in a certain requisite manner along the trajectories of the closed system. As in other problems, here we seek the control action in the form of the function of both the output x_A of the system and time.

Assume that we have solved this problem and found a function $v = v(x_A, t)$ such that if

$$x_B = v(x_A, t) \tag{8.8}$$

the quantities x_A and x_B vary in the desired fashion. We can now turn to the problem of the synthesis of a control $u(x, t)$ that causes the system B to operate so as to fulfill equality (8.8). If this equality is valid, it will be said that the control $u = u(x, t)$ induces internal feedback with the operator $v(x_A, t)$ (more exactly, induces it at the zero error). It is evident that the synthesis problem for an inducing control (the induction problem, for short) is the tracking problem: the output x_B of the dynamic systemm B must reproduce the function $v(x_A(t), t)$.

Hence, the control synthesis problem for the object (8.3) entails two problems: one of them serves to make up the control for the system A and thus provide the operator of inducible feedback and the other serves to form the control for the system B by inducing internal feedback with the obtained operator. The replacement of the initial problem by the two interrelated problems for control of objects of smaller dimensions lies at the heart of our lecture course.

But it is worthy of note that the solution of the second problem — the induction problem in its preset statement — involves severe and, sometimes, insurmountable difficulties. We introduce the designation $s = x_B - v(x_A, t)$ and call the quantity s the vector of an induction error. What underlies the induction problem is that it is necessary to ensure that s goes to zero in a finite time and subsequently to sustain the exact equality $s = 0$. These strict requirements are not specific to control problems; a more common requirement is that a control error signal (equal to the deviation of the controllable quantity from the preassigned value) falls within the neighborhood of zero and, sometimes, asymptotically tends to zero. What causes concern is that it is impossible to form the inducing control in the class of continuous functions under the uncertainty conditions of the object model.

The second-order example considered in Sec. 6 supports this concern. The immediate control aim for the object of the form

$$\dot{x}_1 = x_2$$
$$\dot{x}_2 = g(x, u, t)$$

is just to induce internal feedback $x_2 = -cx_1$. This aim is achieved only for systems with strong feedback and systems with varying structures, i. e., only under the discontinuous control action.

8.4. Induction with Error

To obviate the above difficulties, we weaken the statement of the induction problem so as to bring it to a common tracking problem. We prescribe a number function $\sigma = \sigma(x_A, t) \geqslant 0$ and try to fulfill the following condition in a finite time:

$$\|s\| \leqslant \sigma(x_A, t) \tag{8.9}$$

The quantity σ will be given the name induction error. It will be shown in the subsequent lectures that the problem of error-prone induction lends itself to the solution without going beyond the scope of the class of continuous functions.

However, the weakened requirements on the induction problem raise some questions. Indeed, what does the fulfillment of the condition (8.9) afford? Equality (8.8) defines a specific internal feedback loop which is responsible for the desired character of motions in the system. Inequality (8.9) does not provide for this definiteness. What is then the sense in the error-prone induction? We will show that under certain conditions, inequality (8.9) is consistent with the conversion of the internal feedback operator of the system (8.7) to one of the elements of a set totally comprising the operators that endow the system with approximately identical and, likewise, satisfactory properties.

Inequality (8.9) can be changed for the equality

$$x_B = v(x_A, t) + \mu\sigma(x_A, t) \tag{8.10}$$

which includes the condition $\|\mu\| \leqslant 1$. As a matter of fact, this notation represents the definition of the quantity μ. It is easy to see that the fulfillment of inequality (8.9) corresponds to the application of feedback of the form (8.10) to the system A.

Let us consider a parametric set of the form (8.10), assuming that the parameter $\mu = \mu(t)$ can be any function that varies within a unit ball $\|\mu\| \leqslant 1$. An arbitrary representative of this set causes the system A to behave according to the equation

$$\dot{x}_A = g_A(x_A, v + \mu\sigma, t) \tag{8.11}$$

In order that the behavior of this system should comply with definite requirements, it is sufficient that the function $v = v(x_A, t)$ could serve as a solution of the problem for automatic control of the object (8.11) provided that $\|\mu\| \leqslant 1$ and $\sigma = \sigma(x_A, t)$ is a certain preset function. In this control function, the action $\mu\sigma$ is subject to an additive disturbance v the value (the norm) of which is limited by the known function σ. This disturbance can be taken nonmeasurable: the originating uncertainty will most likely cause no marked complication of the problem for the synthesis of v because, apart from this uncertainty, the right side of (8.11) must include some indefinite parameters "inherited" from the initial model (8.3).

So, freedom in choosing the function σ is only limited by the solvability condition of the control problem for the object (8.11); the function v must determine a control action for this object. In this case, the requirements placed on the function $x_A(t)$ will be met. Besides, account should be taken of the condition (8.10) in order to obtain the desired pattern of changes in $x_B(t)$. Thus, additional requirements, which do not need to be met, are imposed on the functions v and σ. In general, we will not dwell on the compatibility of the above requirements; some considerations of this issue will be given in the text later on.

Note that in the second-order examples presented in Sec. 6, the replacement of the immediate control objective by the inequality $|x_2 + cx_1| \leqslant \Delta|x_1|$ represents the transition to a weaker statement of the induction problem: instead of the zero error, the error $\delta = \Delta |x_1|$ is taken.

8.5. Chains of Induction Problems

In the text presented above we suggested replacing the control problem under study by two problems of smaller dimensions: the problem of the error-prone induction and the problem called upon to shape the operator of inducible internal feedback and also to select the induction error function. The problem of shaping the operator v is the control problem for the object (8.11). If any of the approaches known in the theory of control are inapplicable to this problem and the heuristic method does not afford any solution, then we can do nothing else, but try to apply the approach developed in this book. If (8.11) is the problem involving a direct action exerted on the controllable object, we turn to the text presented in Sec. 5 and obtain the solution. If this is not the case, the last hope for the solution can lie in the use of the text outlined in this section.

Assume that the model of the system (8.11) can lend itself to the same conversions as those of the model of the initial object (8.3) in passing to (8.7). In other words, there is an opportunity to decompose the state vector x_A and control vector v to sums of two projections in such a way that equation (8.11) with the selected induction error function is equivalent to the system (Fig. 8.3)

$$\dot{x}_{AA} = g_{AA}(x_A, v_A, t)$$
$$\dot{x}_{AA} = g_{AB}(x_A, v_A, v_B, t)$$
(8.12)

If there is no way to do this, we dismiss the issue; the problems of this type are beyond the scope of our lecture course.

Suppose that the above presentation exists and, moreover, $v = v_B$, but v_A is lacking. Then, the problem for the synthesis of v breaks up into two problems of smaller dimensions. The action v shows up as an inducing control; the operator of inducible internal feedback and the induction error should be sought by examining the system in the state x_{AA}.

Thus, the initial control problem for the object (8.3) breaks up into three sequential problems of which the two problems are induction ones. Clearly, using this line of reasoning, we can extend such a chain of problems. This chain begins with the control problem for a certain small-dimensional object, the solution of which enables us to form an inducible feedback operator. On solving the induction problem at a certain error, we obtain an operator of the next internal feedback loop, etc. The initial problem must be solved using a certain known method or the means treated in Sec. 5.

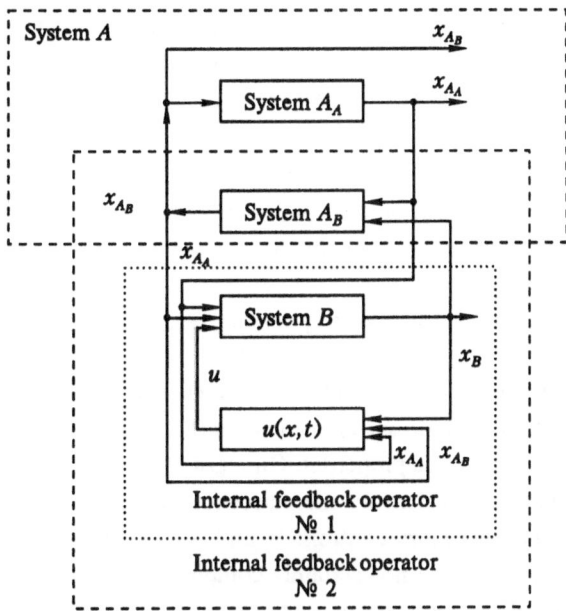

Fig. 8.3. Internal feedback No. 1 induced by the control $u(x,t)$ exerts an action on the system A, which is the subsystem of a controllable object. It is suggested that the objective of this action is to induce some internal feedback (No. 2) in the system A, which is built around a certain subsystem A_A of the system A. Proceeding with the reasoning in a similar way, it is possible to evolve a chain of internal feedback loops that produce one another.

If the initial problem is amenable to the solution in a certain way, there arises a need for synthesizing sequentially an inducing action a few times. Hence, the induction problem serves as the most important element of the approach suggested in this book. Problems of this type are dealt with under additional assumptions in the section that follows.

Note that the total number of problems in one chain is undoubtedly finite because in making up this chain we sequentially pass to the examination of the dynamic systems of progressively smaller dimensions. The number of these systems cannot exceed the dimension of an initial object.

8.6. Clusters of Induction Problems

Such a simple chain of the induction problems as that examined above results from the fact that we impose appreciable constraints on the class of the objects at hand. We return to Sec. 8.2 and recall that up till now we have studied only the systems which have the form (8.6). Similar constraints are suitable for (8.12) and for all successive small-dimensional problems as well. In this case, the entire vector u acts as one inducing control and the entire operator v of inducible internal feedback does likewise. In the general case, all is not as simple as might appear at first sight.

We will refer to equations (8.7). Let the projection of the control u_B solve an induction problem, i. e. provide the condition of the form (8.9). The system A of the controllable object (8.7) then satisfies the equation

$$\dot{x}_A = g_A(x_A, v + \mu\sigma, u_A, t) \tag{8.13}$$

and the condition $\|\mu\| \leqslant 1$. The functions $\sigma(x_A, t)$, $v(x_A, t)$, and $u_A(x_A, t)$ must be defined so that the solution $x_A(t)$ of equation (8.13) and the quantity x_B (t) of (8.10) can display the desired properties. On registering the function $\sigma(x_A, t)$, we must solve the control problem for the object (8.13), which is similar to (8.11), but the former includes more controls: v and u_A. Considering the condition (8.10), the control u_A need not certainly comply with the requirement for the desired character of variations in the vector x_B.

What is the way of controlling the object (8.13)? Let us simplify the notation of the object model

$$\dot{x}_A = \tilde{g}_A(x_A, \omega, t) \tag{8.14}$$

taking ω to be the joint control vector comprising all the coordinates of v and u_A. The problem of the synthesis of ω can be treated in the same way as the control problem of the object (8.3). Here, the vector ω is resolved into a sum of the projections ω_A and ω_B, where ω_B serves as a tool in evolving a certain internal feedback loop in the system (8.14) and ω_A acts on a subsystem of this system, which has the inducible feedback loop placed around it.

Reasoning further in a similar way, we can obtain a set of interrelated induction problems and one initial control problem for the object of the least dimension. The induction problems will be tied together not in a chain, but in a complex cluster: the solution (i. e. the inducing control vector) of each of these problems can contain coordinates of the inducible feedback operators of a few other problems and coordinates of the control u itself. The number of problems in a cluster is finite, as it is in a chain, and these problems need be solved in a definite sequence.

When we mentally take a look at the huge cluster of interrelated induction problems, a shadow of doubt forces its way into our minds: have we not unduly complicated matters? What justifies the line of reasoning that has led us to an intricate situation?.

8.7. What is the Need for so Many Problems?

We will outline the essence of the doubts that appear in the simplest case of the chain of two induction problems including, of course, one initial problem. In this case, a control action forces the subsystem B of the initial system to produce an effect on the system A in such a way that B causes a certain subsystem A_B of the system A to act on the other subsystem A_A of the system A in a definite manner. Assume that B is a system composed of subsystems B and A_B. This designation is likely to indicate that the length of the chain decreases by one link. Really, the control u now makes the system $\tilde{B}$ influence A_A in the required way. It is obvious that in this way we can reduce a longer chain to the chain containing only one induction problem. A similar way can be used to simplify the clusters of problems.

Continuing on the same lines, we can omit the only induction problem left in the chain, considering that the control u just acts on the object, i.e., precisely does what the problem stipulates. Hence, we can return to the starting point of our discourse. The initial problem appears more appropriate than the cluster of problems: the thing is that this problem is only one problem of this type, and although its dimension is comparatively large, this drawback becomes of no concern if we calculate the total dimension of all problems in the cluster.

The range of our reasoning closes on itself due to the following error. We initially considered the technique of control synthesis, i.e., reflected on the issue of how to determine the function $u(x,t)$. The cluster (the chain, in particular) of interrelated induction problems was the result of our reflections. We took the position of the creator of a control, then imperceptibly moved to the position of the user, and began to ponder over how the construction of the control $u(x,t)$ acts on a controllable object. It is quite natural that here the cluster of problems, which were set up with much difficulty, lost any meaning: the controllable object, i.e., the user of the function $u(x,t)$, is completely indifferent to troubles that the creator of this function takes. In the strict sense, the technique of the synthesis of $u(x,t)$ has some meaning for the user only because this technique has an effect on the properties of the function $u(x,t)$. We have not yet touched on the aspects of this type.

Thus, clusters of problems on the whole can be thought of as proving themselves in the right: what matters is that they help the creator of the function. The question that still remains open is what clusters help the creator in each specific case and what kind of help they give. We will take the following line of reasoning. The interrelation between the problems in one cluster, complex as it may be, is strictly defined by the cluster structure and, hence, is conceptual. It is essential that each individual problem in the cluster should lend itself to a solution to enable us to derive the function $u(x,t)$.

As noted above, the induction problems belong to the class of tracking problems for indefinite nonlinear vector objects. What tools do we have at our disposal for the solution of problems of this type? Only the results outlined in Sec. 5. Therefore, in shaping up the cluster of induction problems, one should strive to state them so that each problem can be an element of the class discussed in Sec. 5. If this cluster cannot be built up, the approach developed in this lecture course is inapplicable to induction problems.

If a cluster of the problems of the above type exists, it is quite often found that such a cluster is not the only one. Different clusters entail different control actions and are responsible for different properties of closed systems. The comparison of the results obtained in this case will be made in the text presented below, but now we should study an individual induction problem, which is dealt with in the succeeding sections of this chapter.

9. Synthesis of the Induction Control

An induction problem is a tracking problem, and so the synthesis of an inducing control must essentially be carried out in the same way as the synthesis of an "ordinary" control in a tracking system. The desire to consider indefinite nonlinear objects compels us to refer to the topics outlined in Sec. 5. However, there is no way of applying mechanically the results presented in Sec. 5 to the induction problem. The questions that arise here, their discussion, and the solution of the problem constitute the subject of study in this section.

9.1. Statement of the Induction Problem

Let us examine the object (8.6). The induction problem calls for constructing the control $u = u(x, t)$ so that the condition

$$\|x_B - v(x_A, t)\| \leqslant \sigma(x_A, t) \tag{9.1}$$

should be valid for a closed system, where v and σ are preassigned vector and scalar functions, respectively, and $\sigma(x_A, t) \geqslant 0$ at all $x_{A'}t$. Figure 9.1 displays diagrams that illustrate an induction problem. We assume that these functions on any bounded set satisfy the Lipschitz condition. In the next section, we will impose some additional constraints on v and σ, which stem from the requirements on the correctness of the closed system.

To use the results presented in Sec. 5, we take it that the vector x_B obeys the differential equation considered in this section. This means that our object is

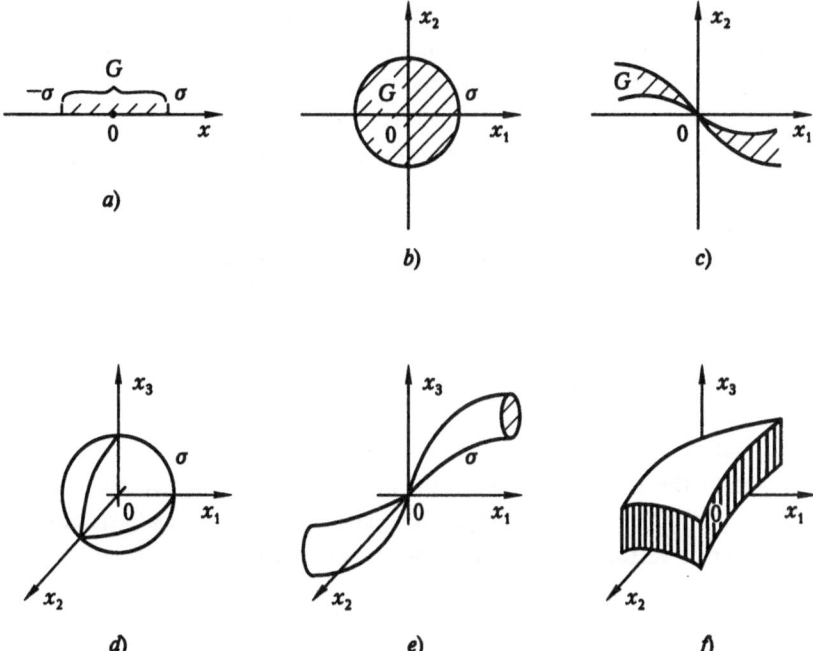

Fig. 9.1. The induction problem involves a definite constraint to be imposed on coordinates of an object. The domain G in the state space corresponds to this constraint; the control must transfer the point x to G and hold it there later on. The domain G represents the neighborhood of the graph of the operator of inducible internal feedback. In view of the nonstationary state, this domain can move and deform during the process of control. The figure displays "instantaneous patterns" of the domains of this kind for objects in which n is the dimension of the vector x and m is the dimension of the vector u: (a) $n = m = 1$; (b) $n = m = 2$; (c) $n = 2$, $m = 1$; (d) $n = m = 3$; (e) $n = 3$, $m = 2$; (f) $n = 3$, $m = 1$.

described by the model

$$\dot{x}_A = \varphi_A(x, t)$$
$$\dot{x}_B = \varphi_B(x, t) + B(t)u \qquad (9.2)$$

where $B(t)$ is the diagonal matrix in which the diagonal elements are continuous and vary in the interval $[1/B, B]$, $B > 1$. Let φ_A and φ_B be continuous functions, which are local Lipschitz ones in x. The current values of components of φ_A, φ_B, and B can be unknown, which indicates that nonmeasurable disturbances of different kinds are available.

We will interpret the statement of the problem from the geometric viewpoint. In the space of coordinates of x, the equation $x_B = v(x_A, t)$ defines a certain manifold — a graph of the operator of inducible internal feedback. Generally speaking, this graph is in motion because a value of v is obviously dependent on time. The problem is to bring the system from an arbitrary initial point $x(t_0)$ to this graph and hold it near the graph in a certain domain specified by inequality (9.1). Sections of the domain cut by manifolds $x_A = \text{const}$ represent balls σ of a variable radius, with the center of each ball lying at the moving point $x_B = u$.

It is easily seen that in Sec. 5 we examined the same problem in a specific case where $x_B = x$ and the dimension of x_A degenerates into zero. The general induction problem for the object (9.2) is rather identical in form to the problem examined in Sec. 5. This is the tracking problem stated for the dynamic system

$$\dot{x}_B = \varphi_B(x_A, x_B, t) + B(t)u \qquad (9.3)$$

i.e., for the system with additional state parameters (combined into the vector x). The difference lies in that the system (9.3) must keep track of the variable $v(x_A, t)$ rather than of identical zero, as is done in Sec. 5. This difference is not crucial if the function v is smooth; in this case, we can replace variables: $s = x_B - v(x_A, t)$. Examine the system behavior in the coordinate space of the induction error vector s. Here, there emerges an equation of the form

$$\dot{s} = \varphi_B - \frac{\partial u}{\partial x_A}\varphi_A - \frac{\partial u}{\partial t} + Bu$$

and so the problem completely agrees with that handled in Sec. 5. But if the function v is nondifferentiable, it is impossible to describe changes in the quantity s by a differential equation.

Then, do we need to introduce an additional condition, namely, the smoothness of u, into the problem statement? We will not follow the unreasoned impulse. If we restrict ourselves only to smooth functions v, then, while gaining in one way, we will heavily lose in another. Let us recall what we managed to clarify in Sec. 8: if the initial control problem is solved by means of a chain of induction problems, then the function v, like any inducing control, affords the solution of a certain tracking problem. But the solution of such a problem, even for the simplest objects considered in Sec. 5, does not always leads to smooth controls. Is it worthwhile to strive for the smoothness? If this is the case, we have to change and complicate the procedure of the control synthesis. It is at this high price that we can simplify the description of the system (9.2) in coordinates of the vector s.

However, as will be apparent from the subsequent discussion, the simplified description of the system alleviates only the analysis of equations. It costs too dearly to pay for this by the synthesis complication, the more so, as will be shown later, the analysis of systems with unsmooth functions v is not so much complex as cumbersome. We generally hold the opinion that it is wrong to sacrifice "practice" (synthesis in this case) to the "theory" (analysis), and not only in the field of automatic control.

9.2. Concept of Control Synthesis

We will develop an inducing control for the case where functions and σ are smooth. It will be clear later on that the same control law is suitable for use in the general case, too. In the lectures that follow, this assertion will be proved formally. But now we put forward only some considerations in favor of this law. Recall that the case in point is the control of an indefinite object, i.e., the control which must be equally efficient at different values of the functions φ_A, φ_B, and B, taken from a wide range, at each point x and at each instant of time t. In other words, one and the same control must be suitable for many objects that markedly differ from one another. In this case, one might expect that, first, this control will solve simultaneously a certain set of induction problems closely resembling each other, i.e., afford the fulfillment of inequalities of the form (9.1) for various pairs of the functions v and σ. Second, there is reason to assume that it is not only the function $u(x, t)$ constructed here that will exhibit the stated property, but also all the functions fairly close to it. As a consequence, the control we have constructed can be considered as a slightly "faulty" one that induces some internal feedback with a smooth operator. The error of this feedback is close to the values of the functions v and σ. At the same time, the feedback loop solves the assigned induction problem.

Thus, we now set out to synthesize a control. It is sufficient for us to derive a function $u = u(x, t)$ that enables, on the strength of (9.2), the quantity $\|s\|$ to decrease steadily at $\|s\|$ from its initial value to the current value of σ, $\|s\| \geqslant \sigma$ but causes this quantity to be stable when $\|s\| \leqslant \sigma$. If the functions v and σ are smooth, then it is sufficient for the above process to fulfill the inequality

$$\frac{d}{dt}\|s\| < -\left|\frac{d}{dt}\sigma\right| \tag{9.4}$$

provided that $\|s\| \geqslant \sigma$. In view of (9.2) we have

$$\frac{d}{dt}\|s\| + \left|\frac{d}{dt}\sigma\right| = \left\langle \frac{s}{\|s\|}, \dot{s}\right\rangle + |\dot{\sigma}|$$

$$= \left\langle \frac{s}{\|s\|}, \varphi_B - \frac{\partial v}{\partial x_A}\varphi_A - \frac{\partial v}{\partial t}\right\rangle +$$

$$+ \left|\left\langle \frac{\partial \sigma}{\partial x_A}, \varphi_A\right\rangle + \frac{\partial \sigma}{\partial t}\right| + \left\langle \frac{s}{\|s\|}, Bu\right\rangle \tag{9.5}$$

$$\leqslant \|\varphi_B\| + \left\|\frac{\partial v}{\partial x_A}\right\| \cdot \|\varphi_A\| + \left\|\frac{\partial v}{\partial t}\right\| +$$

$$+ \left\|\frac{\partial \sigma}{\partial x_A}\right\| \cdot \|\varphi_A\| + \left\|\frac{\partial \sigma}{\partial t}\right\| + \left\langle \frac{s}{\|s\|}, Bu\right\rangle$$

To fulfill the condition (9.4), it is sufficient that the last summand in (9.5) should be negative and relatively large in magnitude. Let $u = -s(\|u\|/\|s\|)$. Then,

$$\left\langle \frac{s}{\|s\|}, Bu \right\rangle = \left\langle \frac{s}{\|s\|}, B\frac{s}{\|s\|} \right\rangle \|u\| \leqslant -\frac{1}{B}\|u\|$$

In this case, for (9.4) to be met, it is sufficient to satisfy the inequality

$$\frac{1}{B}\|u\| > \|\varphi_B\| + \left(\left\|\frac{\partial v}{\partial x_A}\right\| + \left\|\frac{\partial \sigma}{\partial x_A}\right\| \right) \|\varphi_A\| + \left\|\frac{\partial v}{\partial t}\right\| + \left|\frac{\partial \sigma}{\partial t}\right| \qquad (9.6)$$

Thus, if the norm of the control vector satisfies inequality (9.6) and the direction of this vector is opposite to that of s, then the condition (9.4) is valid.

9.3. Control Law

Let it be known that in the right side of the model (9.2), the function $\Phi = \Phi(x, t)$ majorizes $\|\varphi_A\|$ and $\|\varphi_B\|$ at all x, t:

$$\Phi(x,t) \geqslant \max\{\|\varphi_A(x, t)\|, \|\varphi_B(x, t)\|\}$$

The condition (9.6) then holds at

$$\frac{1}{B}\|u\| > \left(1 + \left\|\frac{\partial v}{\partial x_A}\right\| + \left\|\frac{\partial \sigma}{\partial x_A}\right\| \right) \Phi + \left\|\frac{\partial v}{\partial t}\right\| + \left|\frac{\partial \sigma}{\partial t}\right| \qquad (9.7)$$

We will resort to a stricter inequality and determine number functions, and $V_x(x_A, t)$, $V_t(x_A, t)$, $\Sigma_x(x_A, t)$, $\Sigma_t(x_A, t)$ the values of which at each point x_A, t serve as Lipschitz constants for the functions v and σ in x_A and t, respectively. We introduce the function

$$F(x, t) = k\big[(V_x + \Sigma_x + 1)\Phi + V_t + \Sigma_t\big]$$

It is easy to see that for $k > B$ the condition

$$\|u\| \geqslant F(x, t) \qquad (9.8)$$

is sufficient. This condition offers advantages over (9.7). First, only upper estimates for the smooth functions v and σ are to be found and there is no need to estimate the norms $\partial v/\partial x_A$ of the Jacobi matrix and the norms $\partial v/\partial t$ and $\partial \sigma/\partial x_A$ of vectors; here, exact calculations do not always prove simple. Second, it is possible to construct the function F in the general case when v and σ cannot be smooth functions.

The direction control vector $-s/\|s\|$ is discontinuous on the manifold $s = 0$. Therefore, to avoid the discontinuity of $u(x, t)$, we will require that $\|u\|$ should go to zero. This requirement is most likely to be incompatible with inequality (9.8). But recall that the condition (9.8) was necessary to fulfill inequality (9.8) which must hold only when. Therefore, we can permit ourselves to break down (9.8) at low values of $\|s\|$. Let us define a function $\widetilde{\Psi} : R^2 \to R$ such that

$$\begin{cases} \widetilde{\Psi}(\alpha, \beta) \geqslant 1 & \text{for } |\alpha| \geqslant |\beta| \\ \widetilde{\Psi}(0, \beta) \equiv 0 \end{cases}$$

The following formula represents the control law:

$$u(x,\, t) = -\frac{s}{\|s\|}\, \widetilde{\Psi}(\|s\|,\, \sigma)\, F(x,\, t) \qquad (9.9)$$

We will point out the features of the derived law, which are the most *important* ones from our viewpoint.

1. To set up a curent value of the control action by the law (9.9), we must have at our disposal:

(a) information on the current values of all the coordinates of x, t;

(b) information on the lower bound of diagonal elements of the matrix B;

(c) the possibility of estimating the current values of v, σ, V_x, V_t, Σ_x, Σ_t, Φ.

2. The control is defined by the product of three cofactors:

(a) the direction cofactor $s/\|s\|$ specifying the direction of $u(x,t)$ at each point and at each instant of time toward the graph of the internal feedbacl operator;

(b) the "force" cofactor $F(x,\, t)$ specifying a value of $\|u\|$ that is sufficient to suppress the effect of all components in the right side of the object model, which could counteract our desire to reduce $\|s\|$;

(c) the smoothing cofactor $\widetilde{\Psi}$, which ensures the control continuity as it decreases $\|u\|$ near the discontinuity points of the direction vector; for $\widetilde{\Psi} = \widetilde{\Psi}(\alpha, \beta)$ it is possible to use, for example, one of the functions

$$\widetilde{\Psi} = \frac{2\alpha}{(\alpha + \beta)}, \quad \widetilde{\Psi} = \frac{\sqrt{2}\alpha}{\sqrt{\alpha^2 + \beta^2}}, \quad \widetilde{\Psi} = \min\left\{1,\, \frac{\alpha}{\beta}\right\}$$

9.4. What Else Is to Be Done?

In the strict sense, we could have cut down the preceding two sections, leaving intact only the definition of (9.9) and the description of the functions used in the definition. The remaining consideirations only serve as comments on formula (9.9) and, of course, do not prove anything. We have no reason to claim that the obtained control law solves the stated problem "by definition". What is to be done to impart sense to our "definitions"?

First, we have to clarify when the equations of the closed system (9.2), (9.9) are correct, i. e., have a solution that is unique. It is only under this condition that the estimates used in Sec. 9.2 of the derivatives $\|s\|$ and σ have the meaning for the smooth functions v and σ.

Second, we should investigate solutions of the closed system (of course, if it is correct) in the genral case when v and σ are not necessarily differentiable. The estimates of derivates must obviously give way to the estimates of increments.

In the section that follows we will center entirely on the study of the properties of the function $u(x,t)$ defined by (9.9) and derive correctness conditions for a closed system.

10. Correctness of the Closed System

We need to deduce the conditions under which the equations for a controllable object subjected to control actions in any of its initial states have unique solutions. For this, proceeding from the assumptions made above, it is sufficient to ensure that the evolved control as a function of the coordinates x and t should be continuous and, moreover, should satisfy the Lipschitz condition along the coordinates x on any bounded set. This section deals with the study of both the continuity and the Lipschitz property of the control (9.9). The subject of study in this section can be taken as the subject that affords an explanatory exercise relating to the course in mathematical analysis. The reader can consult Sec. 10.3 which presents the summary of the obtained results.

10.1. Continuity of the Control

The function $u(x,t)$ defined by (9.9) is continuous at points where the cofactors $s/\|s\|$, $\widetilde{\Psi}(\|s\|,\sigma)$, and $F(x,t)$ are continuous. The cofactor $F(x,t)$ can be thought of as continuous everywhere because it is in our power to bring it to the desired form. The control proves continuous wherever the product of its other two cofactors is continuous.

The direction vector $-s/\|s\|$ is continuous where $s \neq 0$. The smoothing cofactor $\widetilde{\Psi}$ can be considered continuous as well.

In the examples suggested above, the function $\widetilde{\Psi}$ also displays this property. But the direction vector undergoes a discontinuity where $s = 0$. We introduced the cofactor $\widetilde{\Psi}$ just to ensure the control continuity at these points. This cofactor causes u to go to zero at $s = 0$ and enables the control to take some small intermediate values in the σ-neighborhood of the manifold: $s = 0$. It is possible to avoid the discontinuity if $\sigma > 0$. But if $\sigma = 0$, the function $\widetilde{\Psi}$ has no opportunity to gain its objective: in the coordinate space of the vector s, the range of intermediate values of the control contracts into the point. Note that the formal estimation attests to a discontinuity of $\widetilde{\Psi}(\alpha, \beta)$ at the point $\alpha = \beta = 0$.

Thus, only the points where $\|s\| = \sigma = 0$ do deserve special consideration. Here, the product $(s\widetilde{\Psi})/\|s\|$ is subject to the discontinuity for certain. The only hope of avoiding the discontinuity stems from the function $F(x, t)$ alone because it must go to zero. But it is impossible to set this function at zero anywhere because it must be consistent with the forms of functions described in Sec. 9.3. Where can its zeros lie? By definition, $0 \leqslant \Phi(x, t) \leqslant F(x, t)$. Therefore, the zeros of F have to be zeros of Φ. But the zeros of F, according to the definition of this function, serve as singular points of the system (9.2) at $u = 0$. On account of the uncertainty of values of the components φ_A and φ_B, we cannot know much of these singular points. We assume that our knowledge of the functions φ_A and φ_B is only sufficient to derive the majorants of their norms $\Phi(x, t)$, which can or cannot satisfy the condition $\Phi(0, t) \equiv 0$. This condition can be taken to mean that the free term in the right side is absent. In this connection, if the functions v and σ are subject to requisite constraints, the identity $F(0, t) \equiv 0$ can exist. We will use only this identity.

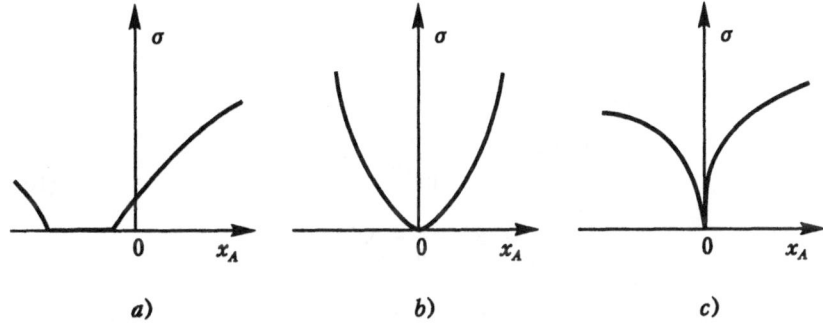

Fig. 10.1. Requirements of the correctness of equations for a closed system impose constraints on the statements of induction problems, i.e., on the set of pairs of the functions v and σ. This figure presents examples of the incorrect assignment of induction errors. The thing is that (a) the function σ can go to zero only when $x = 0$; (b) the fraction $\|x_A\|/\sigma$ must be bounded; and (c) the function σ must satisfy the Lipschitz condition.

According to the adopted assumptions, the control (9.9) is continuous under the condition $\|s\| = \sigma = 0$ if only the coordinate origin in the space of the point x conforms to this condition. Then, $F(0, t) = 0$ and the discontinuity of the function $\widetilde{\Psi}(\|s\|, \sigma)$ at this point is not infinite. Therefore, to afford the continuity, we should set up the implication (Fig. 10.1)

$$\sigma(x_A, t) = 0 \Rightarrow \|x_A\| + \|v(x_A, t)\| + F(0, t) \equiv 0$$

Besides, we should take it that the function $F(x, t)$ is continuous everywhere and the function $\widetilde{\Psi}(\alpha, \beta)$ is continuous at $|\alpha| + |\beta| > 0$ and bounded in a certain neighborhood of zero. In what follows, we assume that these conditions are valid.

10.2. The Lipschitz Condition

We will clarify in what cases an increment of the control on any bounded set of variations in the coordinates of x, t has the asymptotics

$$\|u(x^*, t) - u(x, t)\| = O(\|x^* - x\|)$$

in going from a certain point x to the nearest point x^* at a prescribed value of t. Here, the constant 0 depends on the choice of the set. Let us simplify the notation by applying the symbols with and without the asterisk to denote values of the function at the point x^*, t and the point x, t, respectively. For brevity, we will speak of the local property of a certain function, bearing in mind that this function exhibits this property on any bounded set of changes in its arguments. We obtain

$$
\begin{aligned}
\|u^* - u\| = {}& \left\| \frac{s}{\|s\|} \widetilde{\Psi} F - \frac{s^*}{\|s^*\|} \widetilde{\Psi}^* f^* \right\| \\
\leq {}& \left\| \frac{s}{\|s\|} \widetilde{\Psi} F - \frac{s}{\|s\|} \widetilde{\Psi} F^* \right\| + \left\| \frac{s}{\|s\|} \widetilde{\Psi} F^* - \frac{s}{\|s\|} \widetilde{\Psi}^* F^* \right\| + \\
& + \left\| \frac{s}{\|s\|} \widetilde{\Psi}^* F^* - \frac{s}{\|s^*\|} \widetilde{\Psi}^* F^* \right\| + \\
& + \left\| \frac{s}{\|s^*\|} \widetilde{\Psi}^* F^* - \frac{s}{\|s^*\|} \widetilde{\Psi}^* F^* \right\| \\
= {}& \widetilde{\Psi} |F - F^*| + F^* |\widetilde{\Psi} - \widetilde{\Psi}^*| + \\
& + \frac{F^* \widetilde{\Psi}^*}{\|s^*\|} \|s - s^*\| + \frac{F^* \Psi^*}{\|s^*\|} \left| \|s^*\| - \|s\| \right|
\end{aligned}
\tag{10.1}
$$

Let us find an upper estimate of each of the four summands in (10.1).

If the function $\widetilde{\Psi}(\alpha, \beta)$ is bounded on any bounded set and the function $F(x, t)$ is a local Lipschitz one in x, then the first summand has a local estimate $O(\|x^* - x\|)$. We accept these properties for $\widetilde{\Psi}$ and F. Note that the function $\widetilde{\Psi}$ is bounded on a compact set on which it is continuous. Therefore, we can demand of this function only its boundedness in a certain neighborhood of the point $\alpha = \beta = 0$. But this assumption has been made in Sec. 10.1, and so the new constraint is only the requirement for the local Lipschitz property of F.

The second summand of (10.1) has the desired estimate where the function $\widetilde{\Psi}(\alpha, \beta)$ is the Lipschitz one in its arguments. Let us require that this condition be locally valid at $|\alpha| + |\beta| > 0$. Then, in the system where throughout, the second summand of (10.1) does not exceed $O(\|x^* - x\|)$. The case where $\sigma(0, t) \equiv 0$ is more complex. The complexity appears only in a certain neighborhood of zero, but the desired estimate exists at $x = 0$. If $x = 0$ and $\sigma(0, t) \equiv 0$, then $F(0, t) \equiv 0$ and the second summand admits the local estimate

$$F^*|\widetilde{\Psi}^* - \widetilde{\Psi}| \leqslant \|F - F^*\|(\widetilde{\Psi}^* + \widetilde{\Psi}) = O\left(\|x^* - x\|\right)$$

on account of the boundedness of $\widetilde{\Psi}$ and the Lipschitz property of F.

Consider the third and fourth summands of (10.1). Because,

$$\left|\|s\| - \|s^*\|\right| \leqslant \|s - s^*\| \leqslant \|x_B - x_B^*\| + \|v - v^*\| = O\left(\|x - x^*\|\right)$$

for reasons of the local Lipschitz property of the function v, these summands are sought in the desired way if the function $(F\widetilde{\Psi})/\|s\|$ is locally bounded. We estimate this function in three intervals: at $\|s\| > \sigma$, at $\sigma > $ const > 0, and at all the remaining points. If $\|s\| > \sigma$, then on any bounded set we have

$$F\widetilde{\Psi} = O\left(\|x\|\right), \quad \|x\| \leqslant \|s\| + \|v\| + \|x_A\| = O\left(\|s\| + \|x_A\|\right),$$

$$\frac{1}{\|s\|} = O\left(\frac{1}{\sigma}\right), \quad \frac{F\widetilde{\Psi}}{\|s\|} = O\left(1 + \frac{\|x_A\|}{\sigma}\right)$$

Placing the requirement of the local boundedness on the fraction $\|x_A\|/\sigma(x_A, t)$ yields the desired estimate. If $\sigma > $ const > 0, then from the Lipschitz property of $\widetilde{\Psi}$ it follows that

$$\widetilde{\Psi} = \widetilde{\Psi}\left(\|s\|, \sigma\right) - \widetilde{\Psi}(0, \sigma) = O\left(\|s\|\right)$$

Therefore, in the case of the local boundedness,

$$\frac{F\widetilde{\Psi}}{\|s\|} = O(F) = O(1)$$

It remains for us to consider the case where $\|s\| \leqslant \sigma$ and the magnitude of σ is close to zero. Under the adopted assumptions we have

$$\|x\| = O(\sigma), \quad \frac{\widetilde{\Psi}F}{\|s\|}O\left(\frac{\widetilde{\Psi}s}{\|s\|}\right)$$

We place the requirement of the boundedness on the function $(\widetilde{\Psi}(\alpha, \beta)\beta)/\alpha$ in a certain neighborhood of the point $\alpha = \beta = 0$. In particular, the function $\widetilde{\Psi}$ in all examples considered above complies with this requirement. In the case under study, the function $(F\widetilde{\Psi})/\|s\|$ turns out to be locally bounded.

10.3. Summary of the Results

The subject of study in Sec. 10.1 offers the proof of the following lemma.

Lemma 1. *If the function $F(x, t)$ is continuous, the function $\widetilde{\Psi}(\alpha, \beta)$ is bounded in a certain mneighborhood of zero and continuous at $\|\alpha\| + \|\beta\| > 0$, and there exists an implication*

$$\sigma(x_A, t) = 0 \rightarrow \|x_A\| + \|v(0, t)\| + F(0, t) \equiv 0,$$

then the function $u(x, t)$ defined by (9.9) is continuous.

Section 10.2 contains the proof of the following assertion.

Lemma 2. *If the conditions of Lemma 1 are valid, the function $F(x, t)$ is the local Lipschitz one in x, the function $\|x_A\|/\sigma(x_A, t)$ is locally bounded, the function $\widetilde{\Psi}(\alpha, \beta)$ is the local Lipschitz one at $\|\alpha\|+\|\beta\| > 0$, and the function $(\widetilde{\Psi}(\alpha, \beta)\beta)/\alpha$ is bounded in a certain neighborhood of zero, then the function $u(x, t)$ specified by (9.9) is the local Lipschitz one in x.*

In what follows, the conditions of this lemma will be taken valid.

11. The System with Induced Feedback: General Properties of Trajectories

The constraints introduced above enable us to assert that the system of equations of interest represents the mathematical model of a closed dynamic system rather than just a peculiar symbolic pattern drawn on paper. We will investigate the properties of the dynamic system, i. e., the properties of solutions (trajectories) of the requisite set of differential equations, and describe the conditions under which the control action we have deduced solves the stated induction problem.

The basic content of this section covers the analysis of changes in some nondifferentiable functions. The analysis is made "head-on", as the saying goes, for which reason it is somewhat cumbersome. The summary of the results is given, as before, at the end of the section.

11.1. Variation of an Induction Error

We will examine changes in the quantity $\|s\|$ on account of the system (9.2), (9.9) and seek the conditions under which this quantity decreases in magnitude at a higher rate than the rate of changes in σ. For differentiable functions, inequality (9.4) gives an adequate insight into this effect. The following assertion is valid.

Theorem 1. *If $k > B$ and $-1 \leqslant \eta \leqslant 1$, then the quantity $\|s\| + \eta\sigma$ decreases monotonically and nonstrictly in view of the system wherever $\|s\| \geqslant \sigma$ and $x \neq 0$.*

Proof. Let at the time t the system remains in the domain $\|s\| \geqslant \sigma$, $x \neq 0$. Consider increments in $\|s\|$ and σ on the time interval $(t, t+\tau)$. We get

$$
\begin{aligned}
\|s(t+\tau)\| - \|s(t)\| &= \|x_B(t+\tau) - v(x_A(t_\tau), t+\tau)\| - \\
&\quad - \|x_B(t) - v(x_A(t), t)\| \\
&\leqslant \|x_B(t+\tau) - v(x_A(t), t)\| - \\
&\quad - \|x_B(t) - v(x_A(t), t)\| + \\
&\quad + \|v(x_A(t+\tau), t+\tau) - v(x_A(t), t)\|
\end{aligned}
\tag{11.1}
$$

Let us calculate the last summand, considering that τ is rather small:

$$
\begin{aligned}
\|v(x_A(t+\tau), t+\tau) - v(x_A(t), t)\| &\leqslant \\
&\leqslant \|v(x_A(t+\tau), t+\tau) - v(x_A(t+\tau), t)\| + \\
&\quad + \|v(x_A(t+\tau), t) - v(x_A(t), t)\| \\
&\leqslant V_t(x_A(t+\tau)) + V_x(x_A(t), t)\|x_A(t+\tau) - x_A(t)\| \\
&= V_t(x_A(t), t)\tau + V_x(x_A(t), t)\|\varphi_A(x(t), t)\|\tau + \\
&\quad + O(\|x_A(t+\tau) - x_A(t)\|\tau) + O(\tau) \\
&\leqslant [V_t(x_A(t), t) + V_x(x_A(t), t)\Phi(x(t), t)]\tau + O(\tau)
\end{aligned}
$$

The first two terms in the estimate (11.1) represent an increment over the period $[0,\ ]$ of the function $\|x_B(t+\Theta) - v(x_A(t), t)\|$ of the argument Θ. This function is differentiable with respect to Θ and hence we have

$$
\begin{aligned}
\|x_B(t+\tau) - v(x_A(t), t)\| &- \|x_B(t) - v(x_A(t), t)\| = \\
&= \tau \frac{d}{d\Theta}\|x_B(t+\Theta) - v(x_A(t), t)\|\Big|_{\Theta=0} + O(\tau) \\
&= \tau < \frac{s(t)}{\|s(t)\|},
\end{aligned}
$$

$$
\varphi_B(x(t), t) + B(t)u(x(t), t) > +O(\tau) \leqslant \left[\Phi(x(t), t) - \frac{1}{B}F(x(t), t)\right] + O(\tau)
$$

Note that it is just here that we have resorted to the conditions of the theorem: at $\|s\| \geqslant \sigma$, the value of the quantity $\|u\|$ is fairly high and at $x \neq 0$ the norm is $\|s\| \geqslant \sigma$ in view of Lemma 1. Thus, we obtain the estimate

$$
\|s(t+\tau)\| - \|s(t)\| \leqslant -\left(\frac{k}{B} - 1\right)[(V_x + 1)\Phi + V_t]\tau - \frac{k}{B}[\Sigma_t + \Sigma_x\Phi]\tau + O(\tau)
\tag{11.2}
$$

where values of the functions are found at the time t and at the point $x(t)$. Besides,

$$\left|\sigma\big(x_A(t+\tau),\, t+\tau\big) - \sigma\big(x_A(t),\, t\big)\right| \leqslant [\Sigma_x \Phi + \Sigma_t]\tau + O(\tau)$$

This establishes the assertion of Theorem 1.

Corollary. *The domain* $G = \{x \mid \|s\| \leqslant \sigma\}$ *is invariant.*

It is not yet sufficient to have the above constraints so as to generate requisite internal feedback in the system at a prescribed error. We need to varify when the motion of the system to the domain G, i.e., a decrease in $\|s\|$, causes the point x to penetrate into this domain.

11.2. Induction Condition for Desired Feedback

The following assertion is valid.

Theorem 2. *Let* $M > 0$, $\delta > 0$ *and* $V \geqslant 0$ *be constants such that* $\Phi(x, t) \geqslant M\|x\|$ *at all* x, t *and also* $\|v(x_A, t)\| \leqslant V\|x_A\|$ *at* $\|x_A\| \leqslant \delta$. *Then at* $k > B$, *a solution (a trajectory) of the system* (9.2), (9.9) *either reaches the domain* G *in a finite time as it moves from any initial point or tend to zero.*

Proof. Let us select the following line of reasoning. We first consider the behavior of the quantity

$$h_\alpha = \|s\| + \alpha\big(\|v\| + \|x_A\|\big) \tag{11.3}$$

along trajectories i.e., solutions $x(t)$, passing beyond G and find out at what values of the parameter α this quantity infinitely diminishes in magnitude. Next, we use the estimate

$$\|x\| \leqslant \|x_B - v\| + \|v\| + \|x_A\| \leqslant \sqrt{1 + \frac{1}{\alpha^2}}\, h_\alpha \tag{11.4}$$

and discover at a decrease in h_α that any solution, which totally lies beyond G, localizes after a certain time in a preassigned, arbitrarily small, neighborhood of the coordinate origin, i.e. tends to zero.

From here on, we use everywhere the designation

$$\Omega_\mu = \{x : \|x\| \leqslant \mu\}$$

As follows from the estimate (11.2) used in Sec. 11.1, a value of $h_\alpha(t)$ steadily decreases at any $\alpha \in \big(0,\, (k/B) - 1\big)$, in which case the following inequality holds for a rather low value of $\tau > 0$ at $x(t) \notin G$:

$$h_\alpha(t+\tau) - h_\alpha(t) \leqslant -\frac{1}{k}\left(\frac{k}{B} - 1 - \alpha\right) F\big(x(t),\, t\big)\tau + O(\tau).$$

We register an arbitrary value of s that lies in the above interval. We obtain

$$h_\alpha(t+\tau) - h_\alpha(t) \leqslant -\left(\frac{k}{B} - (\alpha+1)\right) M\|x(t)\|\tau + O(\tau) \qquad (11.5)$$

If $\|x(t)\| > \nu > 0$, i.e., if the system moves beyond Ω_ν at a certain value of $\nu > 0$, then we have

$$h_\alpha(t+\tau) - h_\alpha(t) \leqslant -\frac{1}{2}\left(\frac{k}{B} - (\alpha+1)\right) M\nu\tau,$$

Consequently, h_α decreases faster than the linear function, and if the system does not fall within either G or Ω_ν by the current time t, then

$$h_\alpha(t) \leqslant h_\alpha(t_0) - \frac{1}{2}\left(\frac{k}{B} - (\alpha-1)\right) M\nu(t-t_0).$$

At the time

$$T_\nu = t_0 + \frac{2h_1(t_0)(1+\alpha)}{\left(\dfrac{k}{B} - (\alpha+1)\right) M\nu} \qquad (11.6)$$

the expression in the right side of (11.6) goes to zero. Hence, at a certain instant within the interval $[t_0, T_\nu]$, T_ν, the conditions under which we have derived inequality (11.6) break down: the system gains access either to G or to Ω_ν. The first event meets our needs because the system remains in G forever after $x(t)$ penetrates into this domain.

Let us consider the second event. The choice of the constant $\nu > 0$ is in our hands because it can be taken for granted that a trajectory which does not reach into the domain G falls within a preassigned neighborhood of zero, however small. But this fact does not yet imply that the system tends to zero: as it penetrates into Ω_ν, the point $x(t)$ can in time leave this domain. Let us show that in this case, too, the system can move beyond G within the confines of the neighborhood of zero the radius of which is only a few times larger than a value of ν, and so the radius can be chosen as small as desired.

Let $\nu \leqslant \delta$. In this case, the inequality $\|v\| \leqslant V\|x\|$ holds for Ω_ν. As a result, once the system enters Ω_ν, the following condition is set up:

$$h_\alpha \leqslant (1+\alpha)(1+V)\nu$$

Considering the monotonicity of h_α beyond G, this inequality cannot break down until the system finds its way into G. Taking into account (11.4), we find out that the system does not escape from the domain Ω_η after the time T, where

$$\eta = (1+\alpha)^2(1+V)\frac{\nu}{\alpha}$$

The proper choice of ν permits us to make η as small as one likes. Hence, for any $\eta > 0$ we can define the instant t_η from which the system moves either in G or in Ω_η. The quantity t_η can be defined by resorting to the expression for ν. For this, we express ν in terms of η and substitute the expression thus obtained into the formula for T_ν:

$$t_\eta = t_0 + \frac{2h_1(t_0)(1+\alpha)^3(1+V)}{\left(\dfrac{k}{B} - (\alpha+1)\right)M\alpha\eta} \tag{11.7}$$

Thus, under the conditions of Theorem 2, the deduced control either solves the stated induction problem or ensures the asymptotic motion of $x(t) \to 0$. The discussion of the common issue as to when any of these two effects occur lies outside the scope of this book. We merely note that there are systems which produce only one of these effects and systems which can produce both of these effects depending on the initial point. For example, if the domain G includes a certain fixed neighborhood of zero, the system at hand certainly acquires the desired feedback loop generated in a finite time with the prescribed error because here the motion of $x \to 0$ beyond G is impossible. An alternative effect appears in the class of elementary systems where singular internal feedback need be induced with the zero error: here all solutions, given the nonzero initial conditions, lie beyond the domain G which degenrates into a point.

It is of interest to analyze the motion of a system along trajectories that totally extend beyond G. These trajectories will be called exceptional. They result from the fact that the motion of the point $x(t)$ slows down near the singularity located at the coordinate origin, which simetimes lies at the boundary of G. The system motion along an exceptional trajectory does not conform to the statement of the induction problem. Therefore, the need arises to perform a specific investigation of exceptional trajectories and to determine the conditions under which they can be taken satisfactory. The natural desire to avoid any trouble in performing this investigation impels us to doubt that it is expedient to develop inducing systems in which the exceptional trajectories appear. In the text that follows we first examine (and remove) the doubts involved and then analyze exceptional trajectories.

11.3. Do We Need Exceptional Trajectories?

Exceptional trajectories originate only in inducible feedback systems where $\sigma(0,t) \equiv 0$ (Fig. 11.1). This identity is admissible only if the right sight of the object model does not contain a free term, i.e., if the coordinate origin is a singular point of the system under the zero control. This situation rarely occurs in practice, and so the analysis of exceptional trajectories is of no practical importance.

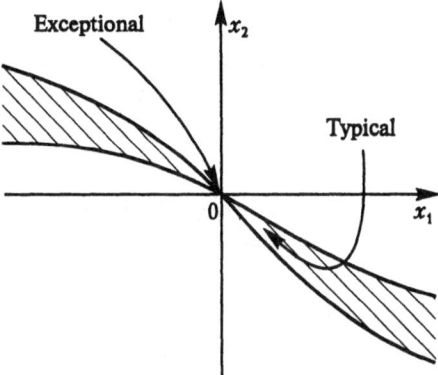

Fig. 11.1. The typical trajectory of an induced feedback system shows that the point x approaches the boundary of G, passes through the boundary, and moves further inside the domain G. The exceptional trajectory corresponds to the case where the coordinate origin lies on the boundary of G and the point $x(t)$ is ready to intersect this boundary just at this site. Exceptional trajectories are met with only in a system that has no free term in the right side of the model. Therefore, their study is of no practical importance.

Is it then worthwhile to take up this analysis? It could be possible to alleviate the task by restricting the discussion only to induction problems in which the function σ does not go to zero anywhere. In this case, apart from the fact that all trajectories would disappear, the correctness analysis (see Sec. 10) would be by far simpler. However, there are two objections to the simpler analysis.

The first objection stems from the conventions of the theory of automatic control. According to this theory, the simplest problem is conventionally held to be the object control problem without the free term (or, as sometimes said, the control problem of a free motion). This viewpoint has obviously come about in connection with the results of the theory of linear controllable systems. The use of conventional methods for the control problems of a free motion makes it possible to ensure the asymptotic stability of the zero solution under definite conditions, i.e., to ensure that $x \rightarrow 0$ under any initial conditions. Can we use our new method to achieve the same for a wider class of objects. The idea of the approach developed in this book lies in the replacement of the initial control problem by a certain induction problem the solution of which serves as a sufficient condition for initiation of the desired motions in the system of interest. But if $\sigma > 0$ everywhere, then no matter how we select the function $v(x_A, t)$, the relation $x \in G$ is not a sufficient condition for $x \rightarrow 0$ to go to zero. If $\sigma(0, t) \equiv 0$, then in some cases the condition $x \in G$ affords the motion of $x \to 0$ when $t \to \infty$; in particular, the example considered in Sec. 6 attests to this fact. So, the analysis of systems under the condition $\sigma(0, t) \equiv 0$ is made in an attempt to avoid the reproach on the part of those who adhere to the conventional methods: if we are given the opportunity to have to do without the free term, we will not be able to make the most of it in a rational way. In fact, we do not let this opportunity slip, although it appears to be an incredible one in the very literal sense: in real life, there is no reason to dream of it.

The second objection against the simplicity relates to the character of the text in itself. We compiled the course of studies in an effort to enable the reader to gain a more complete insight into the approach suggested, where possible. It seems to us that the study of some properties of exceptional trajectories can help the reader to gain a more profound understanding of the details of operation of closed systems, which will also be useful to those who want to be engaged in the practical application of the approach and to those who turn their efforts to theoretical studies in this field.

We will limit the investigation of exceptional trajectories to a problem of deriving the asymptotics of a decrease in $\|x(t)\|$, considering that anyhow that is something if $x(t)$ tends to zero fairly fast.

11.4. Hyperbolic Asymptotics

Some insights into the character of how x tends to zero can be gained from the intermediate results presented in Sec. 11.2. On proving Theorem 2, we have clarified that the point x which moves outside G and penetrates into will remain in the latter domain, at least after the time t_η defined by (11.7). In other words, if $t \geqslant t_\eta$ we have $\|x(t)\| \leqslant \eta$. Consequently,

$$\|x(t)\| \leqslant \frac{2h_1(t_0)(1+\alpha)^3(1+V)}{\left(\dfrac{k}{B} - (\alpha+1)\right) M\alpha(t - t_0)} \tag{11.8}$$

A disadvantage of the estimate (11.8) is that it depends on the parameter α which we have introduced arbitrarily. Prescribing different values of α, we obtain different estimates $\|x(t)\|$ that are all valid. Which of these estimates is the strictest? Differentiating the right side of (11.8) with respect to α, we find that the estimate reaches a maximum at

$$\alpha = \frac{k}{B} - \sqrt{\frac{k^2}{B^2} - \frac{k}{B} + 1}$$

For this reason, the following estimate holds:

$$\|x(t)\| \leqslant \frac{2h_1(t_0)(1+V)}{M(t-t_0)} \cdot \frac{\frac{k}{B}+1-\sqrt{\frac{k^2}{B^2}-\frac{k}{B}+1}}{\left(\sqrt{\frac{k^2}{B^2}-\frac{k}{B}+1}-1\right)\left(\frac{k}{B}-\sqrt{\frac{k^2}{B^2}-\frac{k}{B}+1}\right)}$$

$$\leqslant \frac{2h_1(t_0)(1+V)}{M(t-t_0)} \cdot 4 \left(\frac{3}{\frac{k}{B}-1}+\frac{2}{\left(\frac{k}{B}-1\right)^2}\right)$$

Hence,

$$\|x(t)\| = O\left(\left[\frac{1}{\frac{k}{B}}+\frac{1}{\left(\frac{k}{B}-1\right)^2}\right]\frac{1}{t-t_0}\right) \tag{11.9}$$

where the constant O only depends on M, V, δ, and the initial conditions.

The estimate (11.9) reveals that the speed of motion of $x \to 0$ along exceptional trajectories must increase with k. But the hyperbolic asymptotics with respect to t manifests itself as a very weak result. On the same assumption it is possible to obtain the exponential estimate of the motion of $x \to 0$.

11.5. Exponential Asymptotics

Consider the behavior of the quantity h_α given by (11.3) in the domain $\Omega_\delta \backslash G$ when $\alpha \in \left(0, (k/B)-1\right)$. Here, the inequality

$$h_\alpha \leqslant \sqrt{1+(\alpha+V+\alpha V)^2}\,\|x\|,$$

is valid, and so it follows from (11.5) that at rather small values of $\tau > 0$, we have

$$h_\alpha(t+\tau)-h_\alpha(t) \leqslant -\left(\frac{k}{B(\alpha+1)}-1\right)\frac{MH_\alpha(t)}{V+1}\tau.$$

Let $t\delta$ be an instant of time after which the system at hand does not leave the domain $\Omega_\delta \backslash G$. Then, at $t \geqslant t_\delta$ we obtain

$$h_\alpha(t) \leqslant h_\alpha(t_\delta)\exp\left(-\left[\frac{k}{B(\alpha+1)}-1\right]\frac{M(t-t_\delta)}{V+1}\right)$$

Let us use the inequalities

$$\|x\| \leqslant \frac{1+\alpha}{\alpha} h_\alpha, \quad h_\alpha \leqslant (1+\alpha)(1+V)\|x\|$$

and calculate the estimate

$$\|x(t)\| \leqslant (1+V)\delta \frac{(1+\alpha)^2}{\alpha} \exp\left(-\left[\frac{k}{B(\alpha+1)} - 1\right] \frac{M(t-t_\delta)}{V+1}\right) \qquad (11.10)$$

The obtained result cannot suit us for the same reason as that by which we earlier rejected (11.8): the asymptotics must not depend on a working parameter. In the parametric set of estimates of the form (11.10), a stricter estimate at high values of t proves to be the one to which there corresponds a small value of α; at low values of t (close to t_δ), to this type of estimate there corresponds a high value of α. We will construct the envelope of lower estimates of the form (11.10). The differentiation with respect to α points out that at each t, a minimum estimate of the form (11.10) conforms to a value of

$$\alpha = 2\left[\frac{kM(t-t_\delta)}{B(V+1)} + \sqrt{\left(\frac{kM(t-t_\delta)}{B(V+1)}\right)^2 + 4}\right]^{-1} \qquad (11.11)$$

Note that as t grows from t_δ to $=\infty$, this value decreases from 1 to 0. It is conceivable that somewhere at the initial stage when t is close to t_δ, the value of α defined by (11.11) exceeds $(k/B) - 1$ and our estimates of the decrease in h_α are no longer valid. But inequality (11.10) is certainly valid even at $\alpha > (k/B) - 1$, and so the appearance of high values of α does not lead to errors.

Substituting (11.11) into (11.10) gives the estimate

$$\|x(t)\| = O\left(\left[\left(\frac{k}{B} - 1\right)(t-t_\delta) + 1\right] \exp\left[-\left(\frac{k}{B} - 1\right)\frac{M(t-t_\delta)}{V+1}\right]\right) \qquad (11.12)$$

11.6. Summary of the Results

The qualitative analysis of the equations of a closed system enabled us to prove the following assertions.

Theorem 1. *If $k > B$, the quantity $\|s\|$ steadily decreases in magnitude and does so at a faster rate (nonstrictly) than σ.*

Corollary. *The domain $G = \{x : \|s\| \leqslant \sigma\}$ is invariant.*

Theorem 2. *Assume that the condition of Theorem* 1 *is valid, constants* $M > 0$, $\delta > 0$, *and* $V \geqslant 0$ *exist such that* $\Phi(x, t) \geqslant M\|x\|$, *and the inequality*

$$\|v(x_A, t)\| \leqslant V\|x_A\|$$

is evident from the inequality $\|x_A\| \leqslant \delta$. *Then, an arbitrary trajectory of the closed system is found to lie in the domain* G *after a certain time or the trajectory turns out to be exceptional, i. e., tends to zero beyond* G.

Theorem 3. *If under the conditions of Theorem* 2, *the point* x *moves along an exceptional trajectory and falls within the domain* $\Omega_\delta \backslash G$ *at* $t > t_\delta$, *then the estimate* (11.12) *holds where the constant* O *only depends on* M, V *and* δ.

12. Errors of Synthesis

The function $\Phi(x, t)$ that determines a value (the norm) of the control vector is preset on the basis of prior data on the right side of a controllable object model. In many practical cases, these data are found to be fragmentary and incomplete. Therefore, it is quite natural that there is a risk to make an error during the synthesis of a control and to obtain too small a value of $\|u\|$ at certain values of the arguments. It is clear that the behavior of the closed system so built up will not meet, on the whole, our wishes. How will it behave in the presence of different errors? The answer to this question permits us to estimate the aftereffects of the errors without either understating or overstating their significance. The section deals with this question.

12.1. Errors Localized in a Certain Domain

Assume that in the course of the synthesis of a control, an error has been made such that the quantity $\|u\|$ has proved to be too small in a certain domain Γ of the state space. Let us suppose that there are no errors beyond Γ. Then, the closed system can move within the domain Γ in the most intricate manner, without ensuring for certain a monotonic decrease in $\|s\|$. If $\Gamma \subset G$, a committed error does not in principle entail any changes in the trajectories of the system: the behavior in the domain G is of no importance for us.

It is quite another matter if $\Gamma \supseteq G$. In the domain $\Gamma \subseteq G$, the control does not generally tame the behavior of the system. We will not perform studies of the trajectories in this domain because they require additional information on the right side of the object model. But even without this information we can get an idea of some properties of the system on investigating the motion only beyond $G \cup \Gamma$.

Note that when $\Gamma \supseteq G$ and $\Gamma \cap G \neq \varnothing$, the domain G can lose the invariance property because the point $x(t)$ may go from G to Γ. It is of interest to clarify what domain will be invariant in this case and also in the case where $\Gamma \cap G = \varnothing$. Another way of putting this question is: what domain will bound possible motions of the point $x(t)$ leaving Γ and moving outside $\Gamma \cup G$?

Beyond $\Gamma \cup G$ the control acts in the same way as we conceived it to act in the course of the synthesis. Therefore, there is obviously no monotonic increase in the quantity

$$\tilde{h}_\alpha = \tilde{h}_\alpha(x) = \|s\| + \alpha\|x_A\| \tag{12.1}$$

when $\|\alpha\| \leqslant (k/B) - 1$. Consequently, once it begins to move from an arbitrary point x^* beyond $\Gamma \cup G$, the system cannot go beyond the domain prescribed by the inequality

$$\|s\| \leqslant \inf_{|\alpha| \leqslant \frac{k}{B} - 1} \left(\tilde{h}_\alpha(x_*) - \alpha\|x_A\| \right)$$

The disjunction of the conditions of this type for all $x^* \in \Gamma$ determines the domain which the system cannot leave as it goes out of Γ. Hence, the following set of points becomes invariant:

$$G_* = \bigcap_{x_* \in \Gamma} \left\{ x \mid \|s\| \leqslant \inf_\alpha \left(\tilde{h}_\alpha(x_*) - \alpha\|x_A\| \right) \right\} \cup G \tag{12.2}$$

As can be readily understood, for the system under study, the assertions of Theorems 1–3 are valid if they involve everywhere the domain G^* instead of the domain G.

12.2. Underestimation of Model Parameters

Assume that we underestimated parameters in the right side of the object model when constructing the control, with the result that the function $\Phi(x, t)$ did not become the majorant of $\|\varphi_A\|$ and $\|\varphi_B\|$. It is not inconceivable that the underestimate caused the condition $\Phi(x, t) \leqslant \min \{\|\varphi_A(x, t)\|, \|\varphi_B(x, t)\|\}$ to be valid at all x, t. In this case, the quantities $F(x, t)$ and $\|u\|$ prove too small on the whole. But the underestimation of $\|u\|$ occurs only in a bounded domain under definite conditions, although the function Φ remains too small everywhere. In the case under study, the information presented in Sec. 12.1 can be found useful in investigating the closed system.

What accounts for the fact that the global underestimation of the quantity leads only to the local insufficiency of the control. Let us imagine expansions of the functions $\|\varphi_A\|$ and $\|\varphi_B\|$ into a power series of coordinates of the vector x. The correctly constructed function Φ must majorize the terms of all orders in these expansions. Suppose that a committed error causes Φ to majorize only some of these terms and there are no terms of the highest order among the underestimated ones. Then, the error will affect the quantity Φ everywhere, but will cause qualitative changes in the pattern of trajectories only near the coordinate origin. Indeed, $\|u\|$ has a "make-weight" of order $((k/B) - 1)\,\Phi$, which ensures that $\|s\|$ tends to zero in a correctly developed system (see the proof of Theorem 2); in an error-prone system, the make-weight compensates for a shortage in the quantity Φ owing to the higher-order terms at relatively high values of k and, moreover, is sufficient to reduce $\|s\|$ (Fig. 12.1).

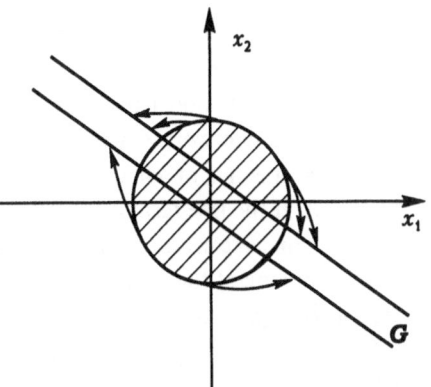

We will not carry out in detail the qualitative analysis of the effect of the compensation for the shortage in lower-order terms owing to the excess of high-order terms and restrict ourselves to a certain procedure of this analysis. For each $\beta \in \left(0, (k/B) - 1\right)$ there exists a radius $r\beta$ of the neighborhood of zero in the space of coordinates of x. Beyond this neighborhood, the quantity $\beta\Phi$ (a portion of the make-weight in the control) exceeds the underesti-

Fig. 12.1. Erroneous notions of object model parameters are apt to lead to an insufficient value of the control in a certain domain Γ (dashed circle). If this is the case, motions of the system in Γ do not comply with the requirements of the task handled. The invariant domain in the closed system then consists of G, Γ, and the domain covered up by trajectories emerging from Γ. The analysis of the size of the third domain enables us to estimate the aftereffects of possible errors made in the procedure of the control synthesis.

mate for the function $\Phi(x, t)$. A value of r_β can be found by comparing a growth of the function $\Phi - \max\{\|\varphi_A\|, \|\varphi_B\|\}$ in $\|x\|$ with a growth of $\beta\Phi$. The quantity $\tilde{h}_\alpha$ defined by (12.1) obviously decreases beyond $\Omega_{r\beta}$ at $|\alpha| < (k/B) - 1 - \beta$. Considering the neighborhood $\Omega_{r\beta}$ to be the domain Γ dealt with in Sec. 12.1, we can reason in a requisite way and find the domain G^* given by (12.2) into which the closed system enters at the end of a certain time when the control process sets in. The size of the domain G^* certainly depends on the magnitude of the parameter β, which we select arbitrarily. To do away with the effect of the arbitrariness on the result obtained, we draw up the diagram of intersections of domains of the G^* type for $\Gamma = \Omega_{r\beta}$ at all possible values of β. The domain so obtained can be thought of as being the result of the deformation of the domain G due to the underestimation of the parameters in the right side of the object model. This deformation lies in that the domain G additionally includes a "bubble" with the center at the coordinate origin. For the object and the function Φ under examination, this bubble must decrease with increasing k.

12.3. Is There Any Need to Find out Errors?

As became evident, an increase in the coefficient k offers the possibility of developing a system that rather weakly responds to an underestimation of a number of object model parameters. In particular, using high values of the coefficient k we can work out controllable systems at $\sigma(0, t) \equiv F(0, t) \equiv 0$ for all objects irrespective of whether or not the free term is present in the right side of the model. If the free term is actually absent, the behavior of the closed system will exactly correlate with our intention. But if this term is available, the system itself will supplement the domain G with the "bubble" near zero, thus sparing us the trouble of replacing the function σ and revising the control law in a requisite way. As a result, we can disregard the fact of whether the free term is present or absent.

Besides, the possibility exists of developing systems at $\sigma(0, t) > 0$ and at a high value of the parameter k, such that the neighborhood $\Omega_{r,\beta}$ beyond which an error does not manifest itself will lie within G. In a system of this type, the underestimation of the model parameters generally does not tell in any way on the character of trajectories: even an error is no hindrance to the system to reach the domain G in a finite time and remain there later on.

Thus, it is possible to evolve "universal" control algorithms: to prescribe a fairly high value of k and hence not to spend additional efforts on the investigation of the object with the aim to avoid making errors. But these "universal" algorithms (systems) are of little practical significance because the effect of the underestimation of the parameters in low-order terms is suppressed through the use of the overestimated parameters in high-order terms which rapidly increase away from zero. The quantity $\|u\|$ then appears to be unwarrantably large at high values of $\|x\|$. The implementation of large control actions may involve severe technical difficulties, which is too high a price for the "universality".

The discussion performed above can virtually be found useful in other respects. In practice, the parameters of an object sometimes behave in the following manner: for the most length of time, they vary within a certain main range, but occasinally increase many times in amplitude over short periods, which gives rise to "bursts" of disturbances. What is the way of majorizing the right sides of the models of such objects? If we account for the range of "bursts", the control will become unwarrantably large in the periods of "calm". But if we only allow for the main range, the control will prove too weak over the "burst" time. In this case, the results presented in Sec. 12.2 and used under certain conditions, will help us to estimate the size of the "bubble" that supplements the domain G and select a value of k such that the size of the bubble and the quantity $\|u\|$ can be not too large.

13. Values of the Control Action in the System with Induced Feedback

To avoid the errors considered in the preceding section, control laws should contain functions that serve for certain as majorants of appropriate quantities. In practice, a majorant has to be defined with a reliability margin which need be larger the lower is the amount of the prior information on an indefinite object. As a result, the quantity $\|u(x, t)\|$ at some x and t (possibly, at all x and t) appreciably exceeds the value sufficient to ensure the desired character of motions of the closed system; the control proves "excessively large" as against the "sufficient" control. In a number of cases, this complicates the technical implementation of the system.

Generally speaking, in developing an inducing control, the designer finds himself at the crossroads since he has to take one of the three ways of the system synthesis, each of which is bound to give him some trouble. First, in the course of the system synthesis, the designer can content himself with comparatively small and readily realizable control actions, taking the risk of committing an error and paying for its aftereffect. Second, the designer can work out the control with a margin so as to avoid errors, but agree with the need to expend additional means for the technical implementation of the system. Third, it is possible to assign additional means for the study of a controllable object in an effort to reduce the indefiniteness of its model and elaborate a comparatively small, but error-free control.

It is possible to compare these ways of the system synthesis only if we introduce additional assumptions as to the cost involved with different troubles. We will not touch on this type of "economy" and restrict our consideration to the study of solutions of the differential systems, which describe the results of the choice of one way or another. We have studied two out of the three ways: one way relates to the error-free systems examined in Sec. 11, in which the value of the control does not arouse interest, and the other relates to the error-prone systems considered in Sec. 12. It remains for us to investigate systems with excessively large-controls, which is the subject of study in this section.

13.1. What Do We Need to Study?

In the text presented below, we perform the comparison between a system with an excessively large control and the same system with a sufficient control for one and the same object. Of course, only one control is generally always available, namely, the control we have worked out. It is of interest to clarify what we lose (or, on the contrary, gain) if our control is in fact too large. To solve this question, we assume that a certain sufficient control, too, is known to us. At first glance, there is no subject for study here. Indeed, as the results presented in Sec. 11 reveal, we can always obtain the required pattern of trajectories when $\|u\|$ at each point is no less than a certain value, so that the excessively large control solves the induction problem as readily as the sufficient control does. Therefore, there is no point in comparing the trajectories, as such, of closed systems. Also, there is almost nothing to say of an individual control action: $\|u(x, t)\|$ at each x, t for the excessively large control exceeds the sufficient level by a value that depends on the margin at which we performed the majorization.

All that holds any interest emerges from the joint consideration of the two issues mentioned above, i. e., from the investigation of changes in the current value of $\|u(t)\|$ along the trajectories of a closed system. The thing is that the function $\|u(t)\|$ is defined by the pair composed of the scalar field of controls and the phase flow of the closed system. Both of the components of this pair depend on the control law, so that the analysis of their joint action represents a nontrivial problem. Our hope of solving it stems from the following simple consideration. The phase velocity of the system with an excessively large control must be fairly high, and thus one might expect that the system will rapidly leave the domain in which the control is too large and will soon pass to the site where the quantity $\|u(x, t)\|$ is rather small, i. e., move closer to the manifold: $s = 0$. As a result, the current value of $\|u(t)\|$ will prove unwarrantably high only on the initial short time interval.

13.2. The One-Dimensional Linear Example

It can be found that an induced control "inherits" the expected effect even from linear systems. Consider the simplest problem of control of the one-dimensional object

$$\dot{x} = f + u \tag{13.1}$$

when $f = \text{const}$. Here, it is necessary to transfer the point x to the prescribed ε-neighborhood of the coordinate origin and not to allow it to escape from this neighborhood. We resort to the proportional controller

$$u(x) = -kx \tag{13.2}$$

The solution of equations (13.1) and (13.2) under the initial condition $x(t) = x_0$ for $t = 0$ is the function

$$x(t) = \left(x_0 - \frac{f}{k}\right) e^{-kt} + \frac{f}{k}$$

Hence, it follows that $x(t) \to f/k$ at $t \to$ and the control (13.2) solves the stated problem at $k > |f|/\varepsilon$. Here, the current value of the control action varies in the following way:

$$u(t) = (f - kx_0)e^{-kt} - f$$

It is obvious that $u(t) \to -f$, in which case

$$|u(t) + f| \leqslant \delta \quad \text{at } t \geqslant \frac{1}{k} \ln \frac{|f - kx_0|}{\delta}$$

Consequently, if the coefficient k in the control law (13.2) is much higher than is necessary, this value of k will affect u only on a small initial time interval which decreases with an increase in k.

13.3. A High Coefficient in the Inducing Control

Passing on to induction problems, we begin with the example similar to the considered example in which the excessively large control is a few times greater everywhere than the sufficient control. Assume that (9.9) specifies the sufficient control satisfying the conditions of Theorems 1 and 2 given in Sec. 11. Let us consider a system with the excessively large control

$$\widetilde{u}(x, t) = -\frac{s}{\|s\|} \widetilde{\Psi}(\|s\|, \sigma) r F(x, t) \tag{13.3}$$

at $r = \text{const} > 1$. This implies that in developing the control, we used a certain function $\widetilde{F}(x, t)$ and now wish to study the consequences of the assumption that it would be possible to apply the function $F(x, t) = (1/r)\widetilde{F}(x, t)$ rather than the function $\widetilde{F}$. In the closed system, the quantity $\|s\| \pm \sigma$ nonstrictly decreases wherever $\|\widetilde{u}\| \geqslant F(x, t)$, i.e., where $\widetilde{\Psi}(\|s\|, \sigma) \geqslant 1/r$. We introduce a function $\sigma_r(x_A, t)$ such that

$$\sigma_r = \max \left\{ \alpha \; \middle| \; \widetilde{\Psi}(\alpha, \sigma) = \frac{1}{r} \right\} \quad \text{at } \sigma > 0 \quad \text{and} \quad \sigma_r = 0 \quad \text{at } \sigma = 0$$

Let

$$G_r = \{x : \|s\| \leqslant \sigma_r\} \tag{13.4}$$

Note that there exists an estimate $\|\widetilde{u}(x, t)\| \leqslant F(x, t)$ in the domain G_r, i.e., the excessively large control in G_r satisfies the same inequality as the sufficient control in G.

We introduce additional constraints under which for a system with the excessively large control (13.3), Theorems 1, 2, and 3 remain valid on replacing the domain G by the domain G_r. We cannot do without additional conditions because $\|s\| - \sigma$ decreases in sigma G_r rather than in $\|s\| - \sigma_r$, so that the domain G_r is not invariant on the whole. For G_r to be invariant, it is sufficient to fulfill the inequality $\|\dot{\sigma}_r\| \leqslant \|\dot{\sigma}\|$. If the function $\widetilde{\Psi}(\alpha, \beta)$ is differentiable at $\|\alpha\| + \|\beta\| > 0$, we have

$$\frac{d}{dt}\widetilde{\Psi}(\sigma_r, \sigma) = \frac{\partial\widetilde{\Psi}}{\partial\sigma_r} \cdot \dot{\sigma}_r + \frac{\partial\widetilde{\Psi}}{\partial\sigma} \cdot \dot{\sigma} = 0$$

Therefore, the domain G_r is invariant if

$$\left|\frac{\partial\widetilde{\Psi}(\alpha, \beta)}{\partial\alpha}\right| > \left|\frac{\partial\widetilde{\Psi}(\alpha, \beta)}{\partial\beta}\right| \quad \text{at } 0 < |\alpha| < |\beta|$$

In the general case where the function is not differentiable, use must be made of Lipschitz constants instead of the absolute values of derivatives. Let us note that the examples of the functions $\widetilde{\Psi}$, which are given in Sec. 9, satisfy the above requirement.

Thus, if the function $\widetilde{\Psi}$ displays the indicated property, the system with the excessively large control either localizes after a certain time in the domain G_r where the function $\|\widetilde{u}\|$ is majorized by the quantity $F(x, t)$ or tends to zero beyond G_r along an exceptional trajectory. The asymptotics of the motion along the exceptional trajectory can be found from the estimate (11.12) by substituting rk for k:

$$x(t) = O\left(\left[\left(\frac{kr}{B} - 1\right)(t - t_\delta) + 1\right]\exp\left[-\left(\frac{kr}{B} - 1\right)\frac{M(t - t_\delta)}{V + 1}\right]\right) \quad (13.5)$$

at $t \geqslant t_\delta$ if the system resides in Ω_δ after t (see the statement of Theorem 3, Sec. 11).

We can by right interpret the inequality $\|\widetilde{u}\| \leqslant F(x, t)$ as an indication of the automatic elimination of the surplus portion in the excessively large control, which distinguishes this control from the sufficient one. This fact is stated more strictly in Sec. 13.1: the control laws (9.9) and (13.3) generate phase flows for which the respective sets G and G_r are invariant; two different pairs in the requisite laws and the invariant sets generate identical estimates of the norm of the control vector.

The estimate $\|\widetilde{u}\| \leqslant F$ does not hold for motions along exceptional trajectories. We can content ourselves with the examination of the asymptotics (13.5) according to which the rate of motion of $x \to 0$ grows with r. Recall that exceptional trajectories are of no practical significance and the studies of excessively large controls are just necessary only for practical purposes. Introducing the condition $\sigma(0, t) > 0$, we can get rid of exceptional trajectories.

Thus, if the algorithm synthesis results in an excessively large control which is many times the sufficient control, the system will automatically reduce the former control by a requisite factor.

13.4. A Varying High Coefficient

We turn to a more common case where the coefficient $r \geqslant 1$ in (13.3) is not constant: $r = r(x, t)$. It is now evidently possible to describe any control that is too large in comparison with the sufficient control (9.9). We consider the domain G_r defined by (13.4). At $r = \text{const}$, the invariance conditions for G_r are found quite natural and not limiting, although this is not the case on the whole. The domain G_r falls within G and contains the manifold $s = 0$. The boundary of G_r can approach either δG at some values of x_A, t or the manifold $s = 0$ at some other values. The domain G_r generally takes the shape and deforms according to the function σ_r which corresponds to the unknown quantity r. In the general case, the invariance of such a "shapeless" set can be ensured only through the use of certain strict and limiting conditions (constraints). We have to admit that the domain G_r is not invariant.

If the point $x(t)$ can move outside the domain G_r, the inequality $\|\tilde{u}\| \leqslant F$ does not hold. To put it otherwise, the system is unable to eliminate automatically the entire surplus portion of the excessively large control. In the case where $r(x, t) \geqslant r_0 = \text{const}$, the condition $\|\tilde{u}\| \leqslant (r(x, t)/r_0)F$ can be expected to set up after some time elapses, which follows from the results presented in Sec. 13.3. This inequality corresponds to the invariance of the domain G_r defined by (13.4), which satisfies the condition $G_r \subset G_{r_0} \subset G$. However, it is not inconceivable that $r_0 = 1$ and $G_{r_0} = G$; in which case the results described in Sec. 13.3 prove ineffective.

We will seek a function $\rho = \rho(x, t)$, $1 \leqslant \rho \leqslant r$ such that the set $G\rho$ (13.4) appears to be invariant and hence $\|\tilde{u}\| \leqslant (r/\rho)F$. Let us clarify when the quantity $\tilde{\Psi}_\rho$ does not increase monotonically in view of the system at $\tilde{\Psi}\rho = 1$. For simplicity, the functions $\tilde{\psi}$, $\|s\|$, σ, and ρ are taken to be differentiable. We obtain

$$\frac{d}{dt}\left(\rho\tilde{\Psi}\right) = \dot{\tilde{\Psi}}\rho + \tilde{\Psi}\dot{\rho} = \left(\frac{\partial\tilde{\Psi}}{\partial\|s\|}\frac{d}{dt}\|s\| + \frac{\partial\tilde{\psi}}{\partial\sigma}\dot{\sigma}\right)\rho + \tilde{\psi}\dot{\rho}$$

Assume, as before, that

$$\left|\frac{\partial\tilde{\Psi}(\alpha, \beta)}{\partial\alpha}\right| > \left|\frac{\partial\tilde{\Psi}(\alpha, \beta)}{\partial\beta}\right| \quad \text{at } 0 < |\alpha| < |\beta|$$

In addition, let

$$\frac{\partial \widetilde{\Psi}(\alpha, \beta)}{\partial \alpha} > 0 \quad \text{when } \alpha > 0$$

The new constraint means that the quantity $\widetilde{\Psi}$ steadily increases with respect to $\|s\|$ at any σ; the quantity $\widetilde{\Psi}$ displays this property in all the examples given in Sec. 9. We derive the inequality

$$\frac{d}{dt}\left(\widetilde{\Psi}\rho\right) \leqslant \frac{\partial \widetilde{\Psi}}{\partial\|s\|}\left[\frac{d}{dt}\|s\| + |\dot{\sigma}|\right]\rho + \widetilde{\Psi}\dot{\rho}$$

From the condition $\widetilde{\Psi}_\rho = 1$ it follows that

$$\frac{d}{dt}\|s\| + |\dot{\sigma}| \leqslant \frac{1}{k}F - \frac{1}{B}\|\widetilde{u}\| \leqslant -\frac{F}{B}\left(\widetilde{\Psi}r - 1\right) = -\frac{F}{B}\left(r - \rho\right)\frac{1}{\rho}$$

Therefore,

$$\frac{d}{dt}\left(\widetilde{\Psi}\rho\right) \leqslant -\frac{\partial \widetilde{\Psi}}{\partial\|s\|}\frac{F}{B}\left(r - \rho\right) + \frac{\dot{\rho}}{\rho}$$

Hence, to determine the invariant domain G_ρ, we should find a solution $\rho(x, t)$ of the inequality

$$\frac{\dot{\rho}}{\rho} - \frac{\partial \widetilde{\Psi}}{\partial\|s\|}\frac{F}{B}\left(r - \rho\right) < 0 \qquad (13.6)$$

The higher the value of $\rho(x, t)$ which we are able to find, the better the result. In the strict sense, the symbol $\dot{\rho}$ identifies the expression

$$\frac{\partial \rho}{\partial t} + \langle \text{grad } \rho(x, t), \dot{x}\rangle,$$

where $\dot{x}$ corresponds to equations of the object model of interest.

Inequality (13.6) confirms our previous conclusion: if there exists a number $r_0 > 0$ such that $r \geqslant r_0$, then $\rho = r_0$ is the solution of this inequality. Besides, we can set $\rho = r$ at the point $r(x, t)$ of a global minimum, i.e., a maximum of $1/r$. It will not be possible to reach any, more profound, conclusions as to the properties of the solutions of (13.6) because not all the problems are virtually solvable. We will content ourselves only with the conclusion that the closed system with the control law (13.3) can automatically eliminate in a finite time a certain fraction of the "surplus" in this excessively large control.

13.5. What is to Be Done
to Effect Savings in Control from the Outset?

The initial section of the trajectory of a closed system can pass where $\widetilde{\Psi} \geqslant 1$ and the excessively large control does not fall off. In this case, there can arise difficulties in implementing the system, at least for a short time. The system will be able to rapidly approach G at the expense of relatively large control actions, although a high speed can prove unnecessary since the pay for it is more than we can afford. Is there any way of abandoning this poor tradeoff? How can we weaken the excessively large control at the initial stage? For simplicity, we restrict the discussion to the case where $r = \text{const}$ in (13.3).

Note that if the initial point $x(t_0)$ lies in G_r, the inequality $\|\widetilde{u}\| \leqslant F$ proves valid at once. The quantity r is essentially unknown, and so we can be sure that $x(t_0) \in G_r$ is true only if the point $x(t_0)$ falls within the manifold $s = 0$. In this case, leaving the above manifold, the system will reside in G_r and the quantity $\|\widetilde{u}\|$, starting at the zero value, will never exceed $F(x, t)$. The following procedure of using this effect suggests itself: we construct a certain temporary manifold $s = 0$ passing through the initial point $x(t_0)$ and use s^* instead of s to derive the control law. Next, once the time $t0$ elapses, we move this manifold until s^* coincides with s. The system will always remain in the domain of small control actions, but after s^* becomes coincident with s, the system will arrive at G where the desired pattern of motions is put into effect.

We will formally describe the advanced proposition. Assume, for example, that s_0 is a value of s at the time t_0, $T > 0$, and

$$s^* = s - s_0 \left(1 - \frac{t - t_0}{T}\right) \quad \text{at } t \leqslant t_0 + T$$

or, otherwise, $s = s^*$. The control then takes the form

$$\widetilde{u} = -\frac{s^*}{\|s^*\|} \widetilde{\Psi}(\|s^*\|, \sigma) \widetilde{F}^*(x, t) \tag{13.7}$$

Note that as compared to $\widetilde{F} = rF$, the quantity $\widetilde{F}^*$ should have an increased value of the norm of the partial derivative with respect to time for the inducible internal feedback operator, for now we use the operator

$$u(x_a, t) + s_0 \left(1 - \frac{t - t_0}{T}\right)$$

instead of $u(x, t)$. This increases $\|\widetilde{u}\|$ at $t < t_0 + T$ by the constant of order $\|s_0\|/T$. We determine T arbitrarily and hence there is reason to believe that a multifold decrease of a control action is important for practical purposes.

Thus, at the initial stage of the control process, $\|\widetilde{u}\|$ exceeds F not by r times, but only by the magnitude of order $\|s_0\|/T$. Being content with this result, we cannot but note some weak aspects of the suggested algorithm. In the system with the control (13.7), the point $x(t)$ has to move at the initial stage near the manifold $s^* = 0$, having no way of either advancing beyond it or turning aside. The control counteracts an increase in s^*, whereas it must only serve to reduce $\|s\|$. Unjustified limitations of freedom, as always and everywhere, lead to waste of control efforts; while the function φ ensures without control a fast decrease in $\|s\|$, the control (13.7) tends to act the other way round. The function φ certainly provides an exceptional benefit; perhaps, in a certain system under certain conditions, the situation will never be so favorable even for an instant. However, the law (13.7) admits no unexpected effects of any kind.

We can suggest another method of reducing excessively large controls at the initial stage, which is free from the above drawback. We will not shift the manifold of zeros of the control to the initial point, but expand it so as to form a domain enclosing this point. For this, we use a new smoothing cofactor in the control law. Let $\widetilde{\Psi}^*(\alpha, \beta, \gamma) = 0$ at $\alpha \leqslant \gamma$ and $\widetilde{\Psi}^*(\alpha, \beta, \gamma) \geqslant 1$ at $\alpha \geqslant \beta + \gamma$. Next, let $\widetilde{\Psi}^*(\alpha, \beta, \gamma)$ take any intermediate values at $\gamma < \alpha < \beta + \gamma$. We set

$$\widetilde{u} = -\frac{s}{\|s\|}\widetilde{\Psi}^*(\|s\|, \sigma, z)\widetilde{F}^*(x, t) \tag{13.8}$$

where

$$z = z(t) = \|s_0\| \max\left\{0, 1 - \frac{t - t_0}{T}\right\}$$

Here too, F^* should include the summand of order s_0/T, namely, the majorant $|\dot{z}|$. In the system with the control (13.8), the point x just resides in the domain where $\|s\| \leqslant \sigma + z$, which contracts into G in the time T. Within the domain $\|s\| \leqslant z$, the control does not preclude motions of the system.

The law (13.8) also allows for the use of functions $z = z(x, t)$ which take the zero values in x and t beyond a certain bounded set (i. e., the functions with a bounded carrier). This makes it necessary to include the majorant $\|\partial z/\partial x\|$ into $\widetilde{F}^*$, but permits us to limit the wandering of the system beyond G within a certain "bubble" of finite dimensions and to exclude the possibility that the system can go into infinity beyond G (recall that nonlinear systems are able to move beyond G in a finite time).

13.6. A Fly in the Ointment

It is obviously possible to suggest many different versions of the control law, which offer savings in control actions starting at the initial time. The choice of a law must conform to specific needs of the task involved. However, it should be kept in mind that all the versions of the control algorithm suffer from a common severe drawback: the current value of the control depends on the coordinates of the initial point. In practice, any control algorithm with the "long-term memo-ry" is unreliable: any random fault that disrupts the continuity of time intervals can entail undesirable aftereffects. For example, assume that a value of the function $\widetilde{F}^*$ in the law (13.8) drops off and remains below the required value for a short period. Perhaps, control signals ceased to arrive at the object, or the value of the function z fell off in a jump-like manner, or something else happened. Before long, the function regains its original value. However, the point x shifted during the fault to a domain where $\|\widetilde{u}\| \approx \widetilde{F}^*$. In this case, the effect of the multifold decrease in $\|\widetilde{u}\|$ vanishes, possibly for a long time, until the system reaches "under its own power" the domain of small values of the control. In essence, once the fault ceases to show itself, the control process sets in again. But at the new initial stage (point), the law (13.8) is unable to offer savings in control actions because it contains the coordinates of the previous initial point.

What if the process should be set to run from the new initial point? But this cannot be done for the reason that we do not know that our control is too large, and so we have the right to interpret the system motion as a consequence of the effect of a certain common disturbance and the approximation of $\|\widetilde{u}\|$ to $\widetilde{F}^*$ as a result of our foresight in regard to savings. If we assume that a certain special indicator of faults is available, the situation improves in a number of cases. But, first, the use of such an indicator is inconsistent with the stipulation we adopted, which prescribes the use of information only on the system coordinates and on the time t. Second, if faults occur rather frequently, the attempts to effect savings in control actions can lead to an increase in $\|s\|$, which is completely inadmissible.

In the text given below, we will not return to the issue of savings in control actions from the initial point in time and will not use algorithms of the forms (13.7) and (13.8) and similar ones.

13.7. Basic Result

The discussion held in Sec. 13.3 offers the proof of the following assertion.

Theorem 4. *Let $\Psi_\alpha(\alpha, \beta)$ and $\Psi_\beta(\alpha, \beta)$ be Lipschitz constants of the function $\widetilde{\Psi}$ at the point α, β in α and β, respectively. If the conditions of Theorem 2 are valid and $\Psi_\alpha(\alpha, \beta) > \Psi_\beta(\alpha, \beta)$ at $0 < |\alpha|, |\beta|$, then the estimate (13.5) where the constant O only depends on M, V, and δ exists for the closed system with the control (13.3), which appears either after a certain time such that $x(t) \in G_r$ and $\|\widetilde{u}\| \leqslant F$ or when $x(t) \to 0$ beyond G_r and after a time $t\delta$ such that $x(t) \in \Omega_\delta$ at $t \geqslant t_\delta$.*

14. Potential Functions of Induced Control

Potential functions find wide use for a vivid description of motions of various dynamic systems. It is the widespread use of the approach involved with these functions that has made us turn our attention to it and outline the obtained results in the context of the potential functions. We do not present here new analytical results, and so the reader can omit this section if he is pressed for time.

The text is meant both for those who are familiar with the subject matter of potential systems and for those who want to familiarize themselves with some of its aspects. In this section, we interpret the above-stated properties of controllable systems. The potential notions are only applicable to some induction systems, but the properties that we want to illustrate are common to all systems.

14.1. Potential Systems.
Levels of Clarity of Representation

The system of the form

$$\dot{\xi} = -\operatorname{grad} H(\xi) \tag{14.1.}$$

is known as a potential (gradient) system and the function $H(\xi)$ is known as a potential function (potential) of this system. Recall that $\operatorname{grad} H$ is the vector composed of partial derivatives $\partial H/\partial \xi_1$, $\partial H/\partial \xi_2$, etc. There are also other designations of the gradient, such as ∇H and $\partial H/\partial \xi$.

As is known, the vector $\operatorname{grad} H$ for any ξ is orthogonal to the line (manifold) of the level of H and points in the direction of an increase in this level. This offers an easy way of obtaining the phase trajectories of the system (14.1) if we prescribe a set of manifolds for the level of the function H in the space of coordinates ξ. We begin with the case where $\xi \in R^2$. Here, the plot of $H = H(\xi)$ represents a surface in R^3, referred to as a potential surface. Let the plane of coordinates ξ be a horizontal one and the axis of H is an upward-directed line. The potential surface then resembles the relief of a land. Let us imagine the following: a plane tangent to the relief at a certain point $(\xi, H(\xi))$, a horizontal plane passing through this point, and a vector lying on the tangential plane, which is normal to the line of intersection of the latter plane with the horizontal plane and points downward. The projection of this vector on the plane of coordinates ξ indicates the direction of $\dot{\xi}$. Assume that a solid ball begins to move over a surface from the point $(\xi, H(\xi))$ and its shadow on the plane of ξ travels along $\dot{\xi}$. Where the surface is locally horizontal the ball remains stationary at a point that conforms to the features of the system.

It is sometimes said that the motion of the system (14.1) is consistent with the motion of the solid ball across the potential surface H. It should be noted that this assertion has a special meaning and cannot be taken literally. In actual fact, the system (14.1) moves in the direction of the resultant of two forces applied to the ball, namely, the gravitational force acting downward in the vertical direction and the support reaction force acting normal to the surface. By the second Newton law, we can accept that the direction of motion of the point ξ corresponds to the direction of acceleration of the ball on the potential surface. Clearly, the speed of the ball depends on its motion at each instant of time. The ball can move in any direction, even uphill. These motions do not any longer relate to the system (14.1). Note that the magnitude of the acceleration does not conform to the quantity $\dot{\xi}$.

Thus, in the context of classical mechanics, the solid ball very poorly illustrates the behavior of the system. To find a way out, we imagine an "anti-Newton" ball that always moves downhile and shifts along the horizontal faster if the hill is steeper. The projection of this peculiar object on the horizontal plane prescribes the motion of ξ.

It is customary to assume that a potential surface offers the clarity of representation of the system behavior. We will not call in question this assumption although we treat the notion of clarity, as shown above, in a specific way.

If a value of the potential function H explicitly depends on time and some other variable parameters, the system motion is obviously consistent with a surface that deforms with time. It is more difficult to extend our notion of clarity to the system (14.1) when $\xi \in R^2$ and $n \neq 2$. The one-dimensional case is trivial (the surface above the plane transforms into a curve above the number axis), but when $n \geqslant 3$, the plot of the function H represents a hypersurface in the space of the dimension in excess of three. Here, geometric patterns, as such, will not do, and so we have to be content with the similarity: constructing mentally three-dimensional patterns, we will adhere to the opinion that the multidimensional objects exhibit much the same properties.

14.2. Construction of the Potential Function of Control

Consider the behavior of a closed system in the space of an induction error vector s (Fig. 14.1). Let the operator $u(x_A, t)$ of inducible internal feedback be a smooth function. We can now smoothly change variables $x_B = s + u(k_A, t)$ in the object model and write the equation

$$\dot{s} = \varphi_s(x_A, s, t) + B(t)u \tag{14.2}$$

where

$$u = -\frac{s}{\|s\|}\tilde{\Psi}(\|s\|, \sigma)F_s(x_A, s, t)$$

Introduce a functional $\|s\|_B = \sqrt{\langle s, B(t)s \rangle}$, which is a well-defined one because $B(t)$ is the diagonal matrix with positive diagonal elements. We represent the control law as

$$u = -\frac{s}{\|s\|_B} \cdot \frac{\|s\|_B}{\|s\|} \tilde{\Psi}(\|s\|, \sigma) F_s(x_A, s, t)$$

Out of all possible functions F_s we can find a function such that

$$u = -\frac{s}{\|s\|_B} \Phi_s(x_A, \|s\|_B, t) \qquad (14.3)$$

What is noteworthy is that we consider only the systems subject to two constraints: first, the function u must be smooth and, second, the control must be representable in the form (14.3). As regards this pracrtical case, we cannot certainly indicate which of the controls suitable for the solution of a control problem admits this representation since the elements of the matrix $B(t)$ are unknown to us. But the law of the form (14.3) exists and we have the right to investigate its effect on an object, knowing that all the remaining laws behave in much the same way. We have

$$\text{grad}_s \|s\|_B = \frac{Bs}{\|s\|_B},$$

$$\Phi_s(x_A, \|s\|_B, t) = \frac{\partial}{\partial\|s\|_B} \int_0^{\|s\|_B} \Phi_s(x_A, \alpha, t)\, d\alpha.$$

Therefore, for the control (14.3) there exists a number function $Q = Q(x_A, s, t)$ (equal to the integral given above) such that

$$\dot{s} = \varphi_s - \text{grad}\, Q \qquad (14.4)$$

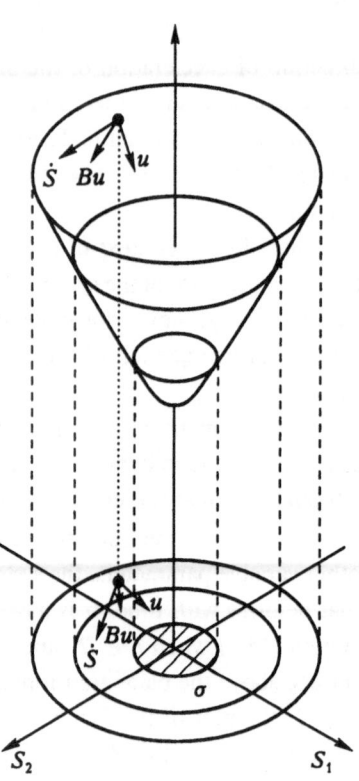

Fig. 14.1. The potential surface, which corresponds to certain induction algorithms, takes the shape of a potential hole with a minimum at the zero point in the coordinates of induction error vectors. If the hole slopes are sufficiently steep, the control affords the requisite speed of the system $\dot{s}$ and hence ensures the motion of s to zero. In a correctly constructed system, the slope steepness is insufficient to produce this effect only within the σ-neighborhood of zero.

Thus, the contribution of the control to the quantity $\dot{s}$ is defined by the potential function.

14.3. Interpretation of Common Properties of Trajectories

The behavior of the system (14.4) can be studied by considering the motion of the above-mentioned "anti-Newton" ball subjected to a disturbance that corresponds to the function φ_s. What is the shape of a potential surface on which the ball rolls? Clearly, the function Φ_s in (14.3) is nonnegative, which goes to zero only at the point $s = 0$. For this reason, the function Q attains a global minimum at $s = 0$, but steadily increases with $\|s\|_B$. Consequently, the potential surface, i.e., the plot of Q, is a potential hole that has a minimun at the point $s = 0$. The shape of this hole varies with time because the quantity Q depends on x_A and t at each s.

Where the slopes of the potential hole are rather steep, the system moves down the slope despite the effect of a disturbance. This character of motion is made feasible at $\|s\| \geqslant \sigma$. Moreover, at $\|s\| = 0$, the hole slopes are steep enough for the ball to roll downward at a higher speed than the speed at which the boundary of the σ-neighborhood of zero is able to approach zero. Thus, the system continues to roll down to the bottom of the hole until it reaches the σ-neighborhood of zero, following which the system is unable to leave this neighborhood. Within the σ-neighborhood, the steepness of the slopes is not sufficient to counteract disturbances φ_s, but now there is no need for a higher steepness. A system without the free term in the right side, for which $\sigma(0, t) \equiv 0$ is valid, can be subject to "collapse" involving the contraction of the σ-neighborhood to a point. In this case the slope steepness is sufficient to ensure that $s \to 0$.

Note that the shape of the potential hole for each closed system depends on the initial conditions for the coordinates of x_A.

14.4. Aftereffects of Errors and Results
   of Preventive Measures

We will interpret from the potential function viewpoint the behavior of systems in which a value of the control does not reach and, on the contrary, exceeds a certain sufficient level.

Assume that the space of x covers a domain Γ where the quantity $\|Bu\|$ is too small in magnitude as against φ_s. In the space of s, to this domain there corresponds its section that changes the shape and can completely vanish on account of motions of x_A. On the potential surface, to this section there corresponds a domain in which the steepness of slopes is too small to counteract disturbances φ_s. In this gently sloping domain, as in the σ-neighborhood of zero, the quantity $\|s\|$ is not bound to decrease. But in contrast to the σ-neighborhood, the flattened domain is not invariant, so that the system can escape from it. If at all x_A and t, the flattened domain lies on the hole slope away from the σ-

neighborhood, then on leaving the flattened domain and moving below it, the system never returns to this domain. But if at certain x_A and t, the flattened domain lies adjacent to the σ-neighborhood, the system can move from the σ-neighborhood to this domain and then return to the σ-neighborhood again, and so on. It is evident that the path along which the system leaves the flattened domain can run across a "normal" and fairly steep slope. This means that the ivvariant domain in the space of x is greater than the union $\Gamma \cup G$.

Suppose that definite preventive measures are taken to preclude errors, with the result that the control (15.3) becomes too large. This means that either the hole slopes turn out to be steeper than expected (if elements $B(t)$ appear to be unexpectedly large) or the disturbance φ_s proves weaker than we thought it should be. One way or the other, the hole slopes are fairly steep to afford a decrease in $\|s\|$ within the σ-neighborhood as well as beyond it. The flattened domain in which φ_s can exert its effect now remains within the σ-neighborhood. This flattened domain is not invariant, and the system can leave and return to it again by moving along a rather steep path. But this wandering occurs only within the σ-neighborhood of zero because the hole slopes at the boundary of the neighborhood are quite steep and prevent the system from moving in the upward direction.

15. Structure of the Induction System

The performed analysis confirms that algorithms of the form (9.9) are quite suitable for solving induction problems. Since the developed algorithms display requisite functional properties, it is worthwhile to give some attention to their structural features, i.e., to the properties of mappings $(x, t) \to u$ of the form (9.9). In this lecture, we handle this issue on the basis of the language of block diagrams, which is in wide use among cyberneticists. The contents of Secs. 15.1 and 15.2 enable novices to eliminate the "language barrier", and those who get an understanding of block diagrams can pass on at once to Sec. 15.3.

15.1. Language of Block Diagrams

A block diagram is a graphic display of elements in the prescribed set of dynamic systems for the purpose of illustrating how the elements interact with one another and with systems that do not enter into this set (Fig. 15.1). Recall that the dynamic system is a certain "device" that converts input signals to output ones. Based on the language of block diagrams, we can depict this device as a rectangle with the set of incoming and outgoing arrows. If the output of some dynamic system is the input of another system, the block diagram displays this features by means of the arrow stretching from one rectangle to the other. Depicting in this way all the systems of interest and hence displaying their interrelationship, we obtain the block diagram composed of the set of rectangles and arrows. A portion of the arrows obviously represent the interrelationship of the above systems and the rest of the arrows depicts the relation of the set of these systems with the environment. The block-diagram language includes special marks to denote commonly occurring systems. We will touch on this aspect in the text below.

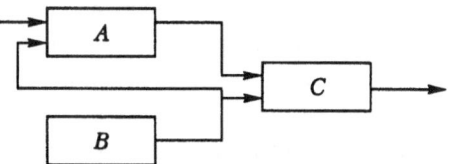

Fig. 15.1. The block diagram is a graphic display of systems of the prescribed set, which shows how they interact with one another and with systems of the outside world. As seen, the inputs of the system C are the outputs of the systems A and B; the output of C exerts a certain effect on the outside world; the output of B is the input of A and simultaneously the input of C; the inputs of B are lacking; the output of A is the input of C; and the inputs of A are the output of B and a certain external action.

The reader familiar with the notions of the theory of graphs can readily reveal the resemblance between a block diagram and an oriented graph. Recall that the oriented graph consists of a set of points called vertices and a set of ordered pairs of these vertices, joined together by arrows called edges or arcs. The block diagram closely resembles the graph, the difference being that the graph cannot contain edges that connect it with the "environment", i. e. the edges extending from its vertices to the environment and the edges coming to the vertices from the environment. It is easy to eliminate this difference if we formally provide the block diagram with a redundant dynamic system corresponding to the environment: this system will absorb signals emerging outward from the original set of dynamic systems and shape incoming signals. But we will not use this approach in our text, nor will we refer to graphs.

Block diagrams serve to depict the interaction of dynamic systems of arbitrary nature. One can imagine a block diagram in which identically shaped rectangles denote, for example, a production unit, an operator of the unit, the environment subject to adverse effects during the production process, and the community that reaps sweet and bitter fruits of industrial development. The arrows can denote the flows of raw materials, finished products, and waste; muscular efforts of the operator, instructions; and so on.

Considering the environment, we single out the dynamic systems and the associated processes which hold interest for the investigation in the case under study. A certain "fragment" of the world, i. e., the environment, can be thought of as different sets of interacting systems according to what problem we are solving. For this reason, the block diagram in question reflects not only the properties of the environment fragment, but the character of the stated problem as well. In general, there is nothing to prevent us from depicting a single rectangle and to take it as a block diagram of the entire world that functions as expected and does not interact with any other dynamic systems because they are nonexistent. This diagram certainly attests to an extreme intelllectural indifference of its compiler. The opposite is evidently inaccessible: any detailed diagram is replaceable by still a more detailed one. The quest for the record-breaking elaboration of a diagram makes no specific sense because all we need to obtain in working out the diagram is just to single out the basic details and leave out the auxiliar ones.

In order to study the behavior of a certain set of dynamic systems, one should examine an aggregate of diagrams of all the individual systems. Of course, purely verbal and qualitative descriptions are not devoid of value, although mathematical models are certainly more preferable. In any event, even the weakest description must contain indications of the input and output of the system of interest. The total aggregate of all indications for the entire set of the systems under study contains all the body of information properly accounted for in the block diagram. Thus, the block diagram is not a supplement to the set of descriptions, but an illustration of the set. The illustration is not a matter of the prime necessity, but there is no point in dismissing it: all the same, the researcher must keep in mind the essence of the interaction of systems, and a better way is to commit it to paper rather than to retain in his memory.

Let us note that use is often made of images that resemble block diagrams, but have quite a different meaning. For example, these images are block diagrams of the algorithms used in programming, diagrams of conversion of substances in the chemical synthesis, and genealogical diagrams (trees) used in paleontology.

15.2. Block Diagrams of Controllable Systems

A controllable system is a set of interacting dynamic systems. Among the processes implemented by this set, two processes are of the utmost importance: the process on which we would like to impose certain constraints and the process that displays our effect on the first one (the control). Two dynamic systems carrying out these processes already afford some rough idea of the controllable system. But it is common to strive to obtain a more detailed representation, for which purpose the treatment is given to a larger number of the processes and the systems involved. The block diagram then becomes more complex, but more useful.

We first turn to the process the course of which we would like to subject to our terms. At the begining of the book, we agreed to limit the range of problems under consideration to tracking problems, i. e. the problems called upon to force a certain controllable quantity to reproduce the current value of the preassigned time function. In this case, the process to be controlled is defined by the deviation of the controllable quantity from the prescribed value, the deviation being called the control error signal. It is necessary to exert an action on the error signal so that its value should rapidly decrease and remain rather close to zero later on. Pondering over this problem, we decided to pass on for a while to a more common question and verify how it is generally possible to cause the error signal to vary in a definite way.

An error signal specifies the process of the joint operation of three interrelated systems: one system, called the setting device, provides a preassigned value of the quantity $y^*(t)$; the other system, called the controllable object or just the object, provides a value of the controllable quantity $y(t)$; and the third sets up the difference between these two values, i. e., forms the error signal $x(t)$. Consider in more detail these three systems.

A value of the controllable quantity $y = y(t)$ is a value of the output of the dynamic system referred to as the controllable object. Two actions serve as inputs of this system: a control action used as an input formed according to our judgement and a disturbance that is an object input resulting from the effect of the environment on the system, which is beyond our control.

The output of the setting device is a vector function $y^* = y^*(t)$. What serves as an input of this device? In general, the quantity $y^*(t)$ can depend on many factors, including the controllable quantity and the control. For example, if y^* is the current market demand for certain goods and y is the curent volume of their production, then y^* can depend on y because the demand decreases in the course of the market saturation. Besides, in some cases there arises the so-called excessive demand if y markedly lags y^*, so that y^* grows still more. The quantity y^* can also increase in magnitude if information on the impending curtailment of production appears. In this case, the dependence of the current value of $y^*(t)$ on the tentative future control will make itself evident. Without ruling out all these effects, we will never account for them and consider a tracking problem on the assumption that the prescribed quantity y^* varies arbitrarily and is not amenable to any effect. This means that we consider the setting device as a dynamic system without inputs.

The system that implements a mapping $(y, y^*) \mapsto x = y - y^*$ is said to be the functional system (the system of functions). Such a "subtractor", along with the adder, is widely met in the field of automatic control. The language of block diagrams prescribes the designation of this system in the form of a circle crossed out with two sloping lines. If the case in point is the adder, the circle displays the shaded sector that corresponds to a deductible signal, and if the system in question performs the addition, all of the four sectors remain unshaded.

It should be stressed that our view of the three interrelated systems that shape up the output $x(t)$ is not devoid of a certain convention. As said above, the block diagram serves as an illustration for the description of a set of dynamic systems. In the case under study, we have only one indeterminate differential equation $\dot{x} = g(x, u, t)$ given in Sec. 3, which describes the only dynamic system with the output $x(t)$ and the inputs, namely, the control u and disturbances. Therefore, the diagram of the three systems is an illustration of something we do not have at our disposal. In the strict sense, this diagram reminds us of the fact that we examine a tracking problem.

We now turn to the diagram that illustrates the shaping of a control. An indifferent researcher will reveal here one dynamic system with the output u. What is the input of this system? We can point up three basic possibilities for the answer, which correspond to three different principles of control. First, the system with the output $u = u(t)$ may not inputs at all. This case conforms to the principle of open-loop control, which underlies the operation, for example, of an automatic washer that can work irrespective of whether or not we loaded it with the washing. Second, a disturbance can serve as an input of the system that operates on the principle of compensating for disturbances. An example of this principle is the supply of workers of certain specialties with additional food to make up for the harmful effect of some production factors on their health.

Third, the input of the dynamic system at hand can be a control error signal x, which complies with the feedback principle. The derived control law corresponds to this control principle.

The control action $u(t)$ shaped up on the feedback principle is the output of the dyanmic system with the input $x(t)$; this system is also known as the feedback operator. The inducing control law (9.9) prescribes the mapping $(x, t) \mapsto u$, and so in our case the feedback operator represents a functional system. Note that this feature is not a necessary one for the inducing control: in Sec. 6 we gave the example of the inducing control operator that is not a functional one. The structural pattern of a control can take the form of a rectangle with the outgoing arrow for u and the incoming arrow for x. The absence of the incoming arrow corresponding to disturbances points out that the information on the disturbances is not used to shape up the control, so that they can be thought of as being nonmeasurable.

The obtained diagram of the controllable system (the object, the setting device, the substractor, and the feedback operator) only points to the fact that we consider the tracking problem and use the feedback principle for its solution. It would be desirable to display in the block diagram the basic features of the derived control law so as to view, as it were, the feedback operator through a microscope. We will examine this issue in Sec. 15.3.

15.3. Structure of the Control Law

The mapping $(x, t) \mapsto u$ which we used in solving the induction problem is described by the formulas presented in Sec. 9. We will draw up a block diagram to illustrate the description of the mapping and single out those features of the suggested control law which, in our opinion, are of the utmost importance for the understanding of the technique of the law deduction.

We take it that the control action is preset by the current value of the function $U\psi(x, t)$ which is an element of the set of the mappings parametrized by the vector Ψ:

$$u = U_\psi(x, t) \tag{15.1}$$

Let

$$U_\psi(x, t) = \Psi \cdot F(x, t) \tag{15.2}$$

Recall that

$$F(x, t) = k\left[(1 + V_x(x_A, t) + \Sigma_x(x_A, t))\Phi(x, t) + V_t(x_A, t) + \Sigma_t(x_A, t)\right]$$

is given by where Ψ stands for majorants of the norms of projections in the right side of the expression; $1/k$ identifies minorants of the nonzero elements of the matrix B; V_x and V_t are majorants of the norms of derivatives of the inducible feedback operator $\sigma(x_A, t)$; and (Σ_x and Σ_t are majorants of the norms of derivatives of the induction error function $\sigma(x_A, t)$.

Hence, at each Ψ the mapping $U_\psi(x, t)$ is found from the prior data on the controllable object, the operator $\sigma(x_A, t)$, and the function $\sigma(x_A, t)$. One and the same set of mappings (15.2) is useful in solving a wide class of induction problems. The limits of this class are obviously preset by the function F the value of which should be high enough to majorize pertinent elements.

The choice of a specific induction problem determines a value of the parameter Ψ. In this connection, we provide for the block diagram an individual operator to establish the mapping $U_\psi(x, t)$ at each Ψ and an individual sequence of operators necessary to shape Ψ.

According to (9.9) and (15.2), the vector ψ is given by

$$\Psi = -\frac{s}{\|s\|}\,\check{\psi}(\|s\|, \sigma) \tag{15.3}$$

We hold that $\Psi = \Psi(s, \sigma)$ is the output of an operator (a system) with the inputs s and σ; $\sigma = \sigma(k_A, t)$ is the output of an operator with the input x_A; $s = x_B - v$ is the output of an operator (a subtrator) with the inputs x_B and v; $u = u(x_A, t)$ is the output of an operator with the input x_A; and x_A and x_B are the outputs of an operator with the input x. The structure of these five operators (systems) specifies the shaping of the signal Ψ that arrives at the operator U_ψ and serves as a parameter.

If we are to hold strictly to theoretic-system ideas, we need to accept that the vector Ψ, like the vector x, is the input of the operator with the output u. But we consider it advisable to stress that Ψ is the parameter of the mapping $U_\psi : (x, t) \mapsto u$. The purpose of the signal Ψ in our system is to select the unique operator from the set of operators of the form (15.2), which corresponds to the induction problem under consideration. In other words, the signal Ψ determines the explicit form of the operator U_Ψ. Hence, the signal U_Ψ will be called the operator or the operator signal and denoted by the thick arrow in the block diagram. All other signals in the system will be referred to as coordinates or coordinate signals. The sequence of conversions of the coordinate x to the operator Ψ will be spoken of as the sequence that results in coordinate-operator feedback and the mapping $U_\psi(x, t)$ of the coordinate x onto the coordinate u with the parameter Ψ will bear the name of coordinate feedback.

Thus, we have completely drawn up the block diagram of the induction system.

15.4. What is a Binary System?

A set of signals that cycle in the developed system admit its subdivision into two subsets: the subset of coordinate signals which are fed to the inputs of the system operators, converted by the operators, and transferred further from their outputs and the subset of operator signals which determine the character of conversions of coordinate signals. The principle of this subdivision is said to be the principle of binarization and the systems that comply with this principle are known as binary systems (Fig. 15.2). Thus, the developed system of the inducing control is a binary system in which the set of operator signals consists of the only signal Ψ.

The principle of binarization, the notion of the operator signal, and the special designation of this signal constitute new elements in the language of block diagrams, which first appeared in the description of dynamic binary systems similar to those given by the example considered in Sec. 6.6. We should now substantiate the introduction of these elements. Note that the question as to whether there is any need for the above elements is not quite correct. Strictly speaking, there is no need to use the language of block diagrams by itself because we can adequately describe the interaction of systems without resorting to this language. Although it is not an obligatory tool, this language is convenient for those who are familiar with it. Cyberneticists use this language in much the same way as doctors use Latin. In general, the language of structures is somewhat similar to any natural language because it serves as a convenient tool in expressing and conveying

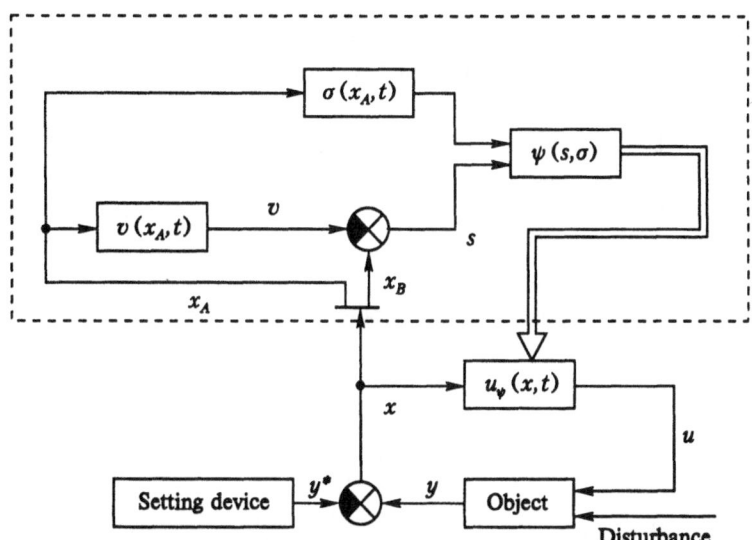

Fig. 15.2. The block diagram of an induction system. A specific designation is used to represent the operator that decomposes the vector x into a sum of two projections x_A and x_B. The thick arrow extends from the operator signal Ψ that determines the explicit form of the mapping $U_\psi : (x, t) \mapsto u$. The dashed line encloses the set of operators intended to generate coordinate-operator feedback.

certain thoughts. To stress the point that some signals in a system prescribe the way of conversion of other signals, we considered it expedient to introduce new elements — neologisms of a sort — into this structure-oriented language. Time will show whether or not they will become current in the language. In our opinion, these neologisms have the right to exist. It may turn out that our opinion is erroneous; as is known from the history of literature, there are cases of the apt and poor creation of words.

Accepting or rejecting the suggested terms, one cannot but notice that systems (binary ones) in which some signals control the conversion of other signals enjoy very wide application. For example, a worker in his factory transforms a workpiece into a finished article, and the properties of this transformation (the speed and quality) are under control of the incoming operator signals, namely, the working conditions and the rate of pay for the work. A green plant converts the solar energy to the energy of organic compounds, and the efficiency of this conversion depends on the environmental conditions which play the part of operator signals. In a computer, the operator signal is a signal that contains the instruction code. The latter example is evidently the most illustrative one in regard to a binary system.

Completing at this stage the discussion of the individual induction problem, we sum up what we have managed to do in this section.

1. Develop the control algorithm and prove that it can generate internal feedback in automatic control systems.

2. Investigate the aftereffects of control synthesis errors and consider the measures taken to prevent these errors.

3. Interpret closed system properties from the potential function viewpoint.

4. Suggest the structural implementation of induction systems relating to the class of automatic binary systems.

We can now turn to the study of the potentialities of induction algorithms for control problems.

Construction of Tracking Systems

The time is ripe to look into the pan set on the gas-stove burner and used to illustrate the procedure outlined in the first section.

The text presented above contains everything that is necessary to develop tracking systems: by the instruction given in Sec. 9, one should interconnect a few induction algorithms set out in Secs. 10 and 11. However, we do not mean to compel the reader to face the need to disentangle on his own the essence of the matter from our text, although the possibility for this is incontestable.

This chapter of the lecture course deals with the procedure of developing controllable systems. Particular attention will be given to the difficulties arising during the procedure and also to the ways of how to obviate them. Clearly, it is by no means possible to resolve all the difficulties since the potentialities of any theory have its limits. The objective of the contents of this chapter is to help the reader to gain a true concept of these limits.

Broadly speaking, to attain this objective, one can consider a number of examples met with in the modern practice of automatic control. We will not follow this path so as not to mislay the theoretical issues under discussion in the host of secondary practical details: as the saying goes, practice is vanity of vanities and anguish of mind. Abstract arguments that the reader will encounter in this chapter are not yet devoid of drawbacks such as cumbersome definitions burdened with the complex notations of variables. But, in our opinion, this does not involve too much trouble.

16. Tracking Systems with Multilevel Binary Structures

As follows from the text set forth in Sec. 8, the chain is the simplest interconnection of induction algorithms that make up a tracking system. In this section, we have to do with the issues of developing the systems of this type and use block diagrams for illustrative purposes.

16.1. Inducing Control in the Tracking System

Consider a control problem for the object

$$\dot{x} = \varphi(x, t) + B(t)u \qquad (16.1)$$

where $x \in R^n$ is a control error vector; $u \in R^n$ is a control vector; $B(t)$ is a diagonal matrix; and $\varphi : R^{n+1} \to R^n$. Assume that we know a constant n_0, $0 < n < n$, such that n_0 diagonal elements of the matrix B are identically zero and the rest of the diagonal elements vary within the section $[1/b, b]$, where $b > 0$. The model (16.1) can now be given as a set of the two vector equations

$$\dot{x}^{00} = \varphi^{00}(x, t)$$
$$\dot{x}^{01} = \varphi^{01}(x, t) + B^1(t)u^1 \qquad (16.2)$$

where $x^{00} \in R^{n_0}$, and $x^{01} \in R^{n-n_0}$. This notation is obviously similar to the notation of (9.2). Why we have considered it necessary to change the designations will be clear from the text that follows.

The components u^1 enter into the right side of the second equation of the system (16.2). Hence, by satisfying formal constraints on the functions v^{01} : $R^{n_0+1} \to R^{n-n_0}$ and $\sigma^{01} : R^{n_0+1} \to R^1$ (see Secs. 10 and 11), we can ensure at the expense of u^1 the generation of internal feedback $x^{01} = v^{01}(x^{00}, t)$ with an error of $\sigma^{01} = \sigma^{01}(x^{00}, t)$. For this, it is just sufficient to know majorants $\|\varphi^{00}\|$ and $\|\varphi^{01}\|$ and minorants of the elements of $B^1(t)$ (the minorants must be positive constants). What form must the functions v^{01} and σ^{01} take in order that the condition

$$\|x^{01} - v^{01}(x^{00}, t)\| \leqslant \sigma^{01}(x^{00}, t) \qquad (16.3)$$

can ensure the desired pattern of motions?

First of all, let us clarify what is meant by the desired pattern of motions. Because the case in point is the tracking problem, we consider it to be our duty to fill one of the two requirements: to ensure the asymptotic stability on the whole of the zero solution with the preassigned exponential estimate of the convergence of transient processes or to secure a fairly fast entry of the point $x(t)$ into the prescribed neihborhood of zero and the localization within this neighborhood of all successive motions of the system. Motions that comply with these requirements are such that the projections x^{00} and x^{01} of the vector x approach zero, each in its own space. The functions v^{01} and σ^{01} must be such that inequality (16.3) should cause the vectors x^{00} and x^{01} to move in the requisite manner.

We write the condition (16.3) in a more convenient form, namely, as the equality

$$x^{01} = v^{01}(x^{00}, t) + \mu \sigma^{01}(x^{00}, t) \qquad (16.4)$$

containing the constraint $\|\mu\| < 1$. As is easy to see, equality (16.4) only asserts that $s^{01} = \mu \sigma^{01}$ and hence provides the estimation of the vector quantity μ. If a value of the function σ^{01} is rather small, then at any $\mu = \mu(t)$ and when $\|\mu\| \leqslant 1$, equality (16.4) implies that the current value of $x^{01} = x^{01}(t)$ is close to $v^{01}(x^{00}, t)$ estimated along the trajectory $x^{00} = x^{00}(t)$. In view of (16.2) and (16.4), the vector x^{00} obeys the equation

$$\dot{x}^{00} = \varphi^{00}(x^{00}, v^{01}(x^{00}, t), t) + \mu \sigma^{01}(x^{00}, t) \qquad (16.5)$$

It still remains for us to estimate the functions v^{01} and σ^{01} in the right side of (16.5). For this, we proceed as follows: first, we prescribe an arbitrary function $\sigma^{01} = \sigma^{01}(x^{00}, t)$, taking into account only the formal constraints on the class of induction error functions presented in Sec. 10, and then solve the tracking problem for the object (16.5), considering that the function v^{01} is a control action and the vector quantity μ is a nonbmeasurable disturbance that varies within a unit ball. Selecting the function σ^{01}, we are generally forced to limit the arbitrariness so that the tracking problem should have a solution.

In constructing the functions v^{01} and σ^{01} we, however, should allow not only for the above factors. For the present, we can only be confident that the condition (16.3) will afford the desired character of motions of the projection x^{00}. The choice of v^{01} should be subject to an additional condition: in view of (16.4), the vector x^{01} must vary along all the trajectories $x^{00}(t)$ of the system (16.5) at any $\mu = \mu(t)$ in a desired way such that $\|\mu\| \leqslant 1$. As a result, inequality (16.3) will become a sufficient condition so as to lead to the desired motions of the point x. It now remains to secure the fulfillment of this condition, i. e., to provide the generation of appropriate internal feedback at a relevant error. This imposes new constraints on the function v^{01} (see Secs. 10 and 11); if v^{01} suits all the imposed requirements, the inducing control $u^1 = u^1(x, t)$ built up by the procedure described in Sec. 9 solves the stated tracking problem for the object (16.1).

16.2. How to State the Induction Problem

The procedure of stating the induction problem of interest involves determining the inducible internal feedback operator $v^{01}(x^{00}, t)$ and induction error function $\sigma^{01}(x^{00}, t)$. In Sec. 16.1 we pointed to the constraints which v^{01} and σ^{01} must satisfy. We now turn to the issue of effective tools for satisfying these constraints, i. e., the issue of how to determine the functions that conform to the introduced requirements.

The essence of requirements consists in the following. The function $v^{01}(x^{00}, t)$ for the selected function $\sigma^{01}(x^{00}, t)$ must handle the task of automatic control of the object (16.5), ensure the desired motions of x^{01} in view of (16.4), and satisfy the formal constraints placed on the inducible internal feedback operator. The function $\sigma^{01}(x^{00}, t)$ must exhibit the formal properties of an induction error and allow for constructing the function $v^{01}(x^{00}, t)$ that satisfies the above requirements. Among the problems corresponding to these constraints, the control problem for the object (16.5) is most important. We now begin to consider this problem.

Any of the properties of the function φ^{00} in the right side of the controllable object model (16.5) can lead to one of the three situations. First, equation (16.5) can belong to the class of equations to which the approach developed in this work is applicable, so that the tracking problem for the model (16.5) can be reduced to a certain induction problem. This case holds the greatest interest and we consider it thoroughly in the next section. Second, the model (16.5) can prove acceptable for any other approaches. Third, the object given by (16.5) can belong to the widest class of objects for which the control methods are not yet known, and so it does not make sense to consider this case.

Let us look at the second case in which the method of synthesis of the control $v^{01}(x^{00}, t)$ for the object given by (16.5) differs in a certain way from our methods. Perhaps, in this case some known methods are suitable, for example, the methods involved in the theory of linear systems, or equation (16.5) can be found simple enough to synthesize v^{01} without resorting to the common method. Here, we also take account of the singular case where $n_0 = 0$, so that the projection x^{00} is absent and equation (16.5) describes an object from the class of elementary objects treated in Secs. 4 and 5, for which it is necessary to generate singular internal feedback $x = 0$. On constructing the function $v^{01}(x^{00}, t)$ in a certain way, we obtain the statement of the induction problem. In general, the constructed pair of the functions v^{01} and σ^{01} may not satisfy the pertinent requirements, and so the stated induction problem may prove unsolvable. It is then worthwhile to seek other functions v^{01} and σ^{01}. If they are unhelpful, we cannot solve the problem, but if they conform to the proper statement of the induction problem, we can develop the inducing control u^1 in keeping with the algorithm (9.9) and hence obtain a precisely identical algorithm, but with other designations.

16.3. Generation of Inducing Internal Feedback

Of primary interest is the case where the control problem (16.5) can admit the same reasoning as the problem (16.1), i. e., when the right side of (16.5) enables us to synthesize $v^{01}(x^{00}, t)$ as an inducing control. For this, the function φ^{00}

should take the form

$$\varphi^{00}(x,t) = \bar{\varphi}^{00}(x,t) + \tilde{B}(t)x^{01}$$

and, moreover, the decomposition of the vector x^{00} into a sum of projections x^{10} and x^{11} should exist so as to write equation (16.5) as a system of the equations

$$\begin{aligned}
\dot{x}^{10} &= \varphi^{10}(x^{00}, t) \\
\dot{x}^{11} &= \varphi^{11}(x^{00}, t) + B^{11}(t)(\mu\sigma^{01}(x^{00}, t) + v^{01}(x^{00}, t))
\end{aligned} \tag{16.6}$$

It is expected that almost all diagonal elements of the matrix $B^{11}(t)$ vary within the section $[1/b, b]$ and the remaining elements are identically zero. If prior information on the object of interest enables us to determine the majorants $\|\varphi^{10}\|$ and $\|\varphi^{11}\|$, the control v^{01} in the system (16.6) allows for solving any properly assigned induction problem no matter what function x^{01} we apply. On stating this problem in the requisite fashion and synthesizing the relevant control action v^{01}, we can ensure that the coordinates x^{00} will vary in the desired manner owing to the fulfillment of inequality (16.3).

It should be noted that the system (16.6) can exist under the adopted assumptions only if the dimension of the vector x^{01} is not higher than that of the projection x^{00}, i.e., if $n_0 \geq n - n_0$. But this is not always the case. So that the results of our discussion should have a large area of application, we take it that a portion of the components of the vector x^{01} serve to impart the desired kind of motions to the projection x^{00} and the rest of the components of x^{01} remain "indifferent". In a similar manner, some of the components of the inducible internal feedback operator $v^{01}(x^{00}, t)$ determine the inducing control for the object with the output x^{00} and the other components of this operator can be set arbitrarily. For the tracking problem at hand, it makes sense to prescribe the latter indifferent components so that they can be identically zero. On substituting these zero components v^{01} in the right side of the equation of the system (16.6), we can undertake to synthesize inducing components. In (16.6), the symbol v^{01} denotes these components. In the strict sense, it would be appropriate to use a new symbol, but we decided to retain the former notation so as not to complicate the description. Note that the outlined method can be used to omit not only "excessive components" of the vector x^{01}, but also those components whose contribution to a value of φ^{00} is nonlinear, or linear, but such that the matrix of appropriate coefficients does not satisfy the requirements placed on B^{11}.

We return to the induction problem related to (16.6). For any function $\sigma^{01}(x^{00}, t)$ we can work out the control $v^{01} = v^{01}(x^{00}, t)$ intended to generate internal feedback of the form $x^{11} = v^{11}(x^{10}, t)$ with an error of $\sigma^{11} = \sigma^{11}(x^{10}, t)$ if the functions v^{11} and σ^{11} satisfy the formal constraints included in the statement of the induction problem. In this case, the function v^{01} has the form similar to (9.9). We should introduce additional constraints under which the above function will display the formal properties of the inducible internal feedback operator. Looking through Secs. 9–11, it is easy to see that now we need to take care of only two properties: the Lipschitz property with respect to t and the boundedness of $\|v^{01}\|/\|x^{00}\|$ within a certain neighborhood of the point $x^{00} = 0$. We put off for a while the discussion of the second property and establish the first by way of the following assertion.

Lemma 3. *If the function $F(x,t)$ is a locally Lipschitz one in t, then this is also the case for $u = u(x,t)$ specified by (9.9).*

We leave out the proof in view of its triviality.

Constructing the control $u^1 = u^1(x,t)$ after the pattern of (9.9), we obtain a means for the generation of internal feedback with the operator v^{01}; this type of feedback in turn serves as a means for the generator v^{11}. When these two identical induction algorithms successively come into action, the coordinates of the system will obey the inequality $\|x^{11} - v^{11}(x^{10},t)\| \leqslant \sigma^{11}(x^{10},t)$. What this leads to depends on how the functions v^{11} and σ^{11} are chosen.

16.4. Construction of the Control System

The problem of building up v^{11} and σ^{11} is a tracking problem for the object

$$\dot{x}^{10} = \varphi^{10}(x^{10}, v^{11} + \mu\sigma^{11}, t)$$

As in the case of the synthesis of v^{01} and σ^{01}, here we can face three situations. First, it can be found that this problem is virtually unsolvable, or unsolvable at the modern stage of development of the theory of control, or solvable, but the derived functions v^{11} and σ^{11} are unsuitable for the correct statement of the induction problem. In these situations, we have to set the initial problem (16.1) aside for the future. Second, the functions v^{11} and σ^{11} that satisfy requisite requirements can sometimes be found without regard for induction systems. On determining these functions, we have stated the induction problem for the object defined by (16.6). Third, the synthesis of v^{11} and σ^{11} can be made in some cases after the pattern of the synthesis of v^{11} and σ^{11}, i.e., on the basis of the approach involved in the lecture course. The vector x^{10} then decomposes into a sum of projections x^{20} and x^{21}. Also, for the selected function σ^{11}, the function v^{11} is built up as an inducing control, i.e., as a means that affords the inequality

$$\|x^{21} - v^{21}(x^{20},t)\| \leqslant \sigma^{21}(x^{20},t)$$

This inequality must secure a definite pattern of motions of the system. For this, the functions v^{21} and σ^{21} need be preset in the appropriate fashion. How can we do this? It is apparent that we should turn to the control problem for the object with the state vector x^{20} and take recourse to the reasoning similar to that carried on above. Perhaps, it will be useful to synthesize the function v^{21} so that it could serve as a means for the generation of internal feedback of the form $x^{31} = v^{31}(x^{30}, t)$ with an error of $\sigma^{31}(x^{30}, t)$. It possibly makes sense to define the pair v^{31}, σ^{31} by applying the suggested approach. Examining the control problems for objects of progressively smaller dimensions, we will finally reach the problem for the synthesis of functions $v^{m_1} = v^{m_1}(x^{m_0}, t)$ and $\sigma^{m_1} = \sigma^{m_1}(x^{m_0}, t)$ which can be set up without using the tools presented in this course. By finding these functions, we can state the induction problem the solution of which will provide the following pair of functions: $v^{m-1,1}(x^{m-1,0}, t)$ and $\sigma^{m-1,1}(x^{m-1,0}, t)$. Finally, we will make up the pair v^{01}, σ^{01} and then the control u^1.

Let us elucidate the essence of the adopted designations. In the suggeste system, the operator $x^{j1} = v^{j1}(x^{j0}, t)$ of internal feedback induced with an error of $\sigma^{j1}(x^{j0}, t)$ serves as a means for the generation of internal feedback of the operator $x^{j+1,1} = v^{j+1,1}(x^{j+1,0}, t)$ with an error of $\sigma^{j+1,1}(x^{j+1,0}, t)$. The vectors $x^{j+1,0}$ and $x^{j+1,1}$ are projections of the vector x^{j0}. Note that the total number of projections x^{01}, x^{11}, $x^{21}, \ldots, x^{m1}$, x^{m0} comprise the vector x, and so our discussion must ultimately come to an end. For now, the second upper index in our designations takes only two values, but in the next section we will come to a more common case where this index will take more values.

Some features of the control synhesis can be clarified by examining the system (16.6) in which the control v^{01} must be built up. The function σ^{01} can be chosen with due regard only for the requirements of the statement of the problem for synthesizing the inducing control u^1. If we do not know that $\varphi(0, t) = 0, \sigma^{01}$ there is a need to prescribe the function σ^{01} at $\sigma^{01}(0, t) > 0$. The right side of (16.6) then contains the free term even if the function $\varphi^{00}(x, t)$ lacks it. Therefore, the function σ^{11} should be set to obey the inequality $\sigma^{11}(0, t) > 0$; the same is also true of σ^{21}, σ^{31}, etc. Besides, as noted above, the function v^1 synthesized on the pattern of the algorithm (9.9) must secure the boundedness of $\|v^{01}\|/\|x^{00}\|$ near the point $x^{00} = 0$. This condition imposes additional requirements on the functions φ^{10} and φ^{11} and also on the pairs v^{11}, σ^{11}, v^{21}, σ^{21}, and so on. It is easy to verify that if the functions v^{m1} and σ^{m1} are Lipschitz ones on any set bounded in x^{m0} and unbounded in t, while values of $\|\varphi^{00}\|$ are bounded at each x, then the above procedure for the construction of a control is feasible. We will not consider any other general cases.

16.5. Multilevel Binary Structure

We will describe binary structures in the language of block diagrams. As noted in Sec. 15, to gain insight into the process of shaping an inducing control, use can be made of a block diagram in which the control action is the output of the operator of coordinate feedback, with the parameter of this operator, or the operator signal, shaped by means of the coordinate-operator feedback loop. This loop includes the inducible internal feedback operator which, as mentioned in the above section, can be built up in the course of solving a certain induction problem. The operator can also be described structurally by including one more coordinate-operator feedback loop in the block diagram.

The structure formed in this way will be called the multilevel structure. The process of developing a control system using structural presentations for its description is as follows. Noting that we will solve the stated tracking problem on the feedback principle, we restrict the discussion to the class of structures in which the control error signal x converts to the control action u (see Fig. 15.3). Accepting that it makes sense to synthesize the mapping $u = u(x, t)$ by the law (9.9) as an inducing control, we turn to binary structures (see Fig. 15.4; let us recall that we changed the designations in this section). In this type of structure, we assign a coordinate-operator feedback loop to the lower structure level. Among the operators of this coordinate-operator feedback loop, there is an inducible internal feedback operator. With the use of the new designations, this operator takes the form of a function $v^{01} = v^{01}(x^{00}, t)$. If the operator v^{01} solves the induction problem for the system (16.6), then in the block diagram we can represent a means of shaping the signal v^{01} as the second coordinate-operator feedback loop at the second structure level. At this level lies one more inducible internal feedback operator that takes the form of a function v^{11}. If this function also plays the role of an inducing control in some small-dimensional dynamic system, the block diagram is made complete with the third structure level, and so on. The process comes to a halt if there is good reason to synthesize at the succesive level the inducible internal feedback operator $v^{m1}(x^{m0}, t)$ without resorting to our approach. Note that the construction of the substructure at the next structure level does not mean that we add some elements to the internal feedback operator illustrated below; on the contrary, we delve deeper and display graphically its design features.

Thus, the sequential examination of induction problems for progressively smaller dimensions is consistent with the sequential development of the substructures at their levels. What is the way of prescribing the operators located at these levels? Let the decision be made to synthesize the next operator $v^{j1}(x^{j0}, t)$ as an inducing control. Then, the signal v^{j1} can be taken as the output of the operator $v^{j1}_{\Psi j+1,1}(x^{j0}, t)$ with the parameter $\Psi^{j+1,1}$, which is built up at the successive level by the coordinate-operator feedback loop. The substructure at this level decomposes x^{j0} into a sum of projections $x^{j+1,0}$ and $x^{j+1,1}$ and implements the mappings

$$(x^{j+1,0}, t) \mapsto v^{j+1,1}, \quad (x^{j+1,1}, t) \mapsto \sigma^{j+1,1}$$
$$(x^{j+1,1}, v^{j+1,1}) \mapsto s^{j+1,1}, \quad (s^{j+1,1}, \sigma^{j+1,1}) \mapsto \Psi^{j+1,1}$$

In any case, even at the stage of the development of the substructure at a definite level, we can specify some of its operators: the operator for the decomposition of the vector x^{j0} into projections; the operator — "subtractor" for shaping $s^{j+1,1}$; and the operator for shaping $\Psi^{j+1,1}$. If the level of interest is the highest one, the operators $v^{j+1,1}$ and $\sigma^{j+1,1}$ are built up by an "extraneous" methods. In particular, it can appear that the synthesis of v^{j1} reduces to the tracking problem for an elementary system, in which case the dimension of the vector $x^{j+1,0}$ degenerates into zero, the quantity $v^{j+1,1}$ is taken to be identically zero, and a value of $\sigma^{j+1,1}$ is found at the required control accuracy along the coordinates of the vector x^{j0}. But if there is reason to continue to build up substructures, then the operator $\sigma^{j+1,1}$ can be defined arbitrarily with the observance of the formal coustraints mentioned above. The determination of the operator $v^{j+1,1}$ has to be set aside because we have need for the information on majorants of the norms of partial derivatives of the as yet unknown function $v^{j+2,1}$. This is not the case for "indifferent" components of the vector function $v^{j+1,1}$ (if they are available), which are not used for the generation of the successive feedback loops: as noted above, they can be preset so as to make them identically zero. Traversing the path upward from level to level, we denote the sites of the operators in the block diagram. Reaching the highest level, we take the route downward across the levels and sequentially fill up empty sites. In this case, we use, among other things, estimates of the norms of the derivatives of $v^{j+1,1}(x^{j+1,0}, t)$ in $x^{j+1,0}$ ut to define, for example, the operator $v^{j1}_{\psi j+1,1}(x^{j0}, t)$.

In brief, the synthesis of a control system is made in two stages: we first take the path from the lowest level upward and sequentially finish building the structure elements at each level and then we follow the path in the downward direction and impart the algorithmic essence to the structure elements.

Figure 16.1 illustrates a two-level binary system; the block diagram at structure level No. $j - 1$ is shown in Fig. 16.2.

16.6. Operation of the Multilevel System

Assume that the construction process described above is complete and the controllable system is put into action. How will it behave? The control u^1 is set up so as to ensure the fulfillment of the inequality $\|x^{01} - v^{01}\| \leqslant \sigma^{01}$ in due course. This problem can sometimes be left unsolved. The control will then

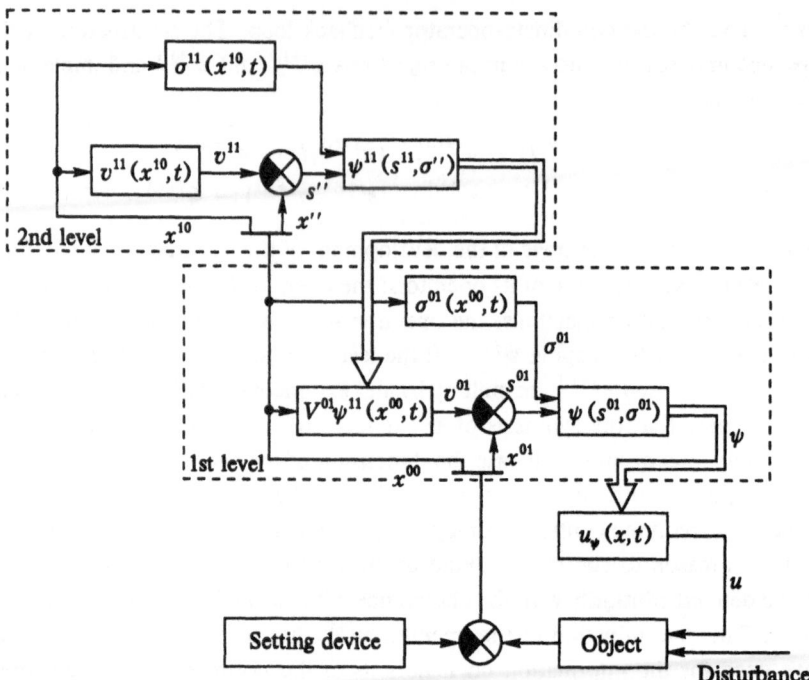

Fig. 16.1. The two-level binary system sequentially generates two internal feedback loops. The binary control $u = u(x,t)$ induces with an error of $\sigma^{01} = \sigma^{01}(x^{00},t)$ the binary internal feedback loop $x^{01} = v^{01}(x^{00},t)$ which, in turn, induces with an error of $\sigma^{11} = v^{11}(x^{10},t)$ the internal feedback loop $x^{11} = v^{01}(x^{10},t)$. Dashed lines enclose the coordinate-operator feedback loops at each of the levels.

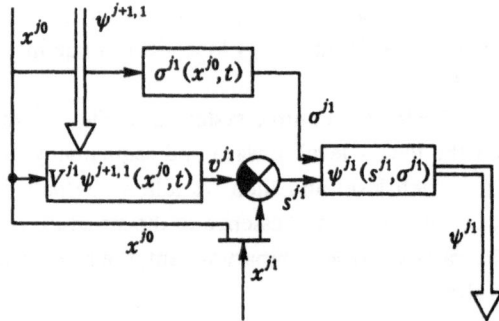

Fig. 16.2. At the structure level No. $j - 1$, the multilevel binary system shapes the operator signal ψ^{j1} which is a parameter of the internal feedback operator at the $(j - 2)$ th level. A similar operator signal $\Psi^{j+1,1}$ is shaped at the jth level if it is available; otherwise, if the $(j - 1)$ th level is the highest, the internal feedback operator $v^{j1}(x^{j0},t)$ is built up at this level without recourse to components of induction systems.

secure the motion of $x \to 0$, which can be taken as a satisfactory alternative under certain conditions. If the above inequality becomes valid, the inducible internal feedback loop with the operator v^{01} will come into play, which either ensures the fulfillment of the condition $\|x^{11} - v^{11}\| \leqslant \sigma^{11}$ or w causes x^{00} to go to zero. The next internal feedback loop will act in a similar way, and so on. Ultimately, the system will be able to satisfy conditions of the form $\|x^{i1} - v^{i1}\| \leqslant \sigma^{i1}$ either for all $i = 0, 1, \ldots, m$ or only for $i = 0, 1, \ldots, j < m$, in the latter case, $x^{j0} \to 0$ when $t \to \infty$.

And what next? Does this result secure required motions such that either all the components of x tend to zero with the desired asymptotics or the vector x rather rapidly falls within the prescribed neighborhood of zero? In the general case, we have to answer in the negative, which is what one should expect: we derived all functions so as to force x^{j0} to move to zero, in which case we disregarded the character of motions of x^{j1}. The cause that led us to this setback is due to the properties of the equations for the object, which may not generally permit us to bring simultaneously all the projections of the vector x closer to zero, whatever control means we apply.

Let us verify this fact using the simplest example of the second-order system described in the coordinates of an error signal by the equations

$$
\begin{aligned}
\dot{x}_1 &= x_2 + 1 \\
\dot{x}_2 &= u
\end{aligned}
\tag{16.7}
$$

Equations (16.7) belong to the class of models for which we can construct a binary system of the type suggested above. But it will not be effective as will not any other control system. Indeed, no matter what controls we use, as soon as we cause the coordinate x_2 to be close to zero, x_1 will begin to increase in view of the second equation of (16.7). The domain enclosing the zero point, in which it is possible to hold the system (16.6), must necessarily intersect the straight line $x_2 = -1$. Therefore, the best that one can do is to select an acceptable domain of this type and leave aside the arbitrarily small neighborhood of zero.

In systems of the general type, as in the case of (16.7), unfavorable effects can appear on account of the free terms present in the equations that lack controls in the right sides. In the notation of (16.2), this means that $\varphi^{00}(0, t) \not\equiv 0$. If this is the case, the stated problem is generally unsolvable; it is necessary to resort to tradeoffs in regard to the equations and weaken the requirements on the problem. The question as to what compromises we should agree to will be left beyond the scope of our course for the reason that it is better to consider each specific case by itself.

We will apply general considerations only to systems where $\varphi^{00}(0, t) \equiv 0$. If the dimension of the vector x^{m0} degenerates into zero at the highest structure level and a value of $\|\varphi^{00}(x,t)\|$ at each value of x is bounded, it is possible to construct a multilevel system that will force the point $x(t)$ to move quite rapidly to the assigned neighborhood of the coordinate origin. If, in addition, $\varphi^{01}(0,t) \equiv 0$, we can ensure that $x \to 0$ will tend to zero with the required asymptotics. Indeed, under the adopted assumption, all the operators of internal feedback loops that generate one another can be preset by the functions the values of which are independent of time, go to zero' at the coordinate origin and, as regards their continuity, are close to zero near the coordinate origin. We will not consider more complex general cases.

In summary, the constructed multilevel binary system can secure the solution of a tracking problem under definite conditions. But the obtained result by no means exhausts the potentialities of the approach under development. In the section that follows we will show how it is possible to weaken these "definite" conditions and to impose more stringent requirements on a closed system.

17. Ways of Extending
the Capabilities of the Approach

In the text presented above, we considered only those controllable systems which involve the simplest form of the interconnection of a few induction algorithms tied together into a chain. In Sec. 8, we described a wider class of the systems in which individual algorithms are tied in clusters. It is reasonable to expect that more complex systems display more extended capabilities. To make sure that this is the case, it would be useful to classify completely the cluster-type systems and describe their capabilities. However, the volume of work that need be done is too large: the list of the equivalence classes of cluster-type systems, for example, from the viewpoint of the theory of graphs, is extremely long and each system has its own functional properties.

For this reason, we will follow another path: indicate a number of effects of practical importance, which are unattainable in chain-type systems, and investigate the possibility of achieving these effects in cluster-type systems. Perhaps, many will not like this practicality, and yet it is typical for cybernetics as well as for any sphere of work where certain means are built up to achieve specific practical ends. For example, the objective in fabricating a hammer is to produce it as a tool for driving in nails, and few care what else it can be used for. So, in this lecture course we leave aside the problem for the complete analysis of the capabilities of cluster-type systems.

The selected line of investigations makes it necessary to take the following initial actions: describe some desired effects, substantiate their practical value, and point out how to achieve them in the class of chain-type systems. These are the issues that we have to do with in this section. As regards the topic of the construction of cluster-type systems for achieving the desired effects, we put it off until the next section.

17.1. Oscillations in Tracking Systems

We will consider an event familiar to cyberneticists as an overcontrol (overshoot). Assume that in the tracking system of interest, the control error signal $x(t)$, taken to be a one-dimensional signal for simplicity, behaves so that $x(t) \to 0$ when $t \to \infty$, the asymptotics of this motion conforming to the requirements of the task involved. The properties of the function $x(t)$ can in no way fix its sign. If the quantity $x(t)$ changes sign in the course of the control process, it is said that an overshoot has occurred: a value of the controllable quantity reaches the prescribed level, goes on rising above this level, and then falls off to it after a time. Following the first overshoot, the quantity may undergo the second, the third overshoot, and so forth. In this case, the motion of $x(t)$ has the form of convergent oscillations.

In general, oscillatory processes occur most widely in nature, society, and engineering (Fig. 17.1). It is possibly this circumstance that impels many practitioners in cybernetics to adnere to the rule that overshoot-prone controllable systems are better than some dubious systems subject to aperiodic motions. In some practical cases, the presence or absence of oscillations is of no convern, but sometimes they prove quite undesirable. To illustrate the point, we can take two examples of routine events: the process of water temperature control in a domestic shower-bath and the relocation of an automobile from row to row on an urban highway. In the first case, the overcontrol at an appreciable amplitude essentially triggers an emotional response, whereas the aftereffects of the overcontrol in the second case happen to be more serious.

In practice, only the oscillations of a relatively high amplitude can pose difficulties, but if the amplitude is small, even continuous oscillations are of no consequence. If the task at hand is to ensure not an asymptotic motion of $x \to 0$, but only the entry of the point x into the preassigned neighborhood of zero, the oscillations within this neighborhood are not certainly taken into account. In this case, appreciable overshoots occur only over a limited time interval after the initial instant of time. This effect, too, as in the examples considered above, may be undesirable.

If an error signal is a vector quantity, then under the term overshoot one understants a change in sign of individual coordinates $x(t)$. As a rule, these coordinates do not change sign at the same time. Of course, one can envision an exotic situation in which the trajectory of a system passes through the origin

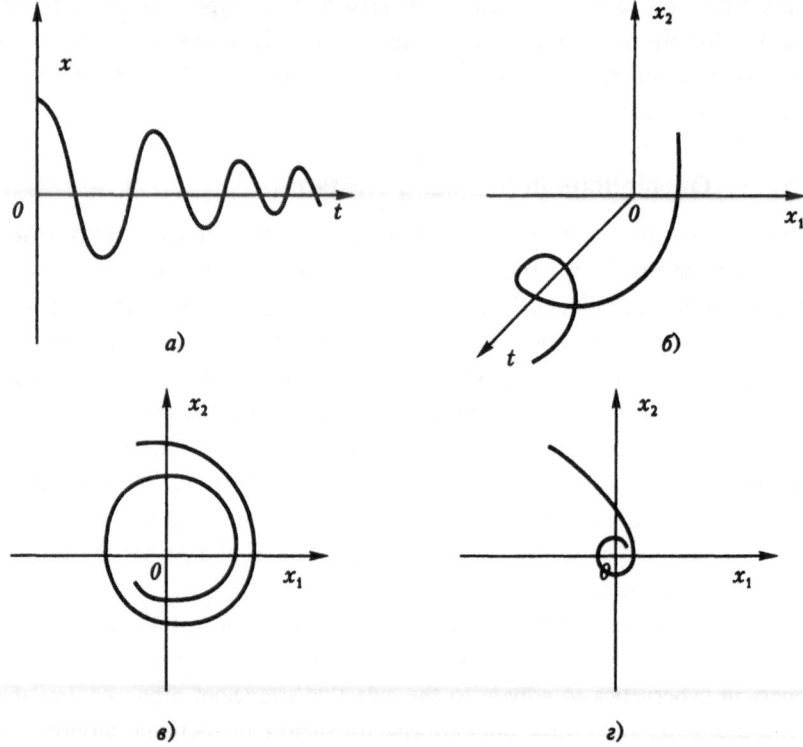

Fig. 17.1. Convergent oscillations in the one-dimensional system (*a*) and two-dimensional system (*b*). The phase portraits (*c*) and (*d*) relate to the second system. An overshoot implies a change in sign of any coordinate of the control error vector. In many practical cases, overshoots are undesirable and are tolerable only if the amplitude of oscillations of a system is sufficiently small, for which reason the phase portrait (*d*) is more preferable than the portrait (*e*).

of coordinates, but a more typical case is the rotation of the system about the coordinate origin. It is clear that the two coordinates change sign simultaneously if the image point of the system in the space of the control error vector passes through a subspace of codimension 2; most likely, the system will "miss its aim". In any event, all such oscillations can lead to undesirable aftereffects. One can submit to these oscillations only if their amplitude is small, namely, the system rotates within a prescribed neighborhood of the coordinate origin, which complies with the required accuracy of control.

17.2. Oscillations in Induction Systems

As is known from above, an inducing control forces the point s in the coordinate space of an induction error signal to approach the coordinate origin until the condition $\|s\| \leqslant \sigma$ sets in, where σ is an induction error. Under certain conditions, the above inequality never holds, but the motion along an exceptional trajectory takes place and $s(t) \to 0$ when $t \to 0\infty$. We assume that motions of systems are satisfactory if the inequality $\|s\| \leqslant \sigma$ holds. Obviously, the point s can oscillate or rotate beyond the σ-neighborhood of zero. If we look at the geometric properties of trajectories, we can reveal approximately the same situation as that considered in Sec. 17.1.

However, there is a difference, which is quite appreciable. In Sec. 17.1 we examined the control error signal whose character is uniquely apparent from the statement of the initial problem and the oscillations of which have a clear practical meaning. A different situation arises with the oscillations of an induction error signal that reflects not the specific feature of the problem involved, but only the technique used for its solution. It is necessary to clarify what role these oscillations and the attendant "regeneration" play in practice.

The easiest way to examine the issue is to consider an example of the elementary system (where $n_0 = 0$ as follows from the notation given in Sec. 16). In this case, the symbols s and x identify the same elements. But if a control action has a smaller dimension than a control error vector, the situation becomes somewhat more complex. Here, to the rotations of s about the σ-neighborhood of zero there correspond rotations of the point x about the domain $G = \{x : \|s\| \leqslant \sigma\}$. These rotations may or may not generate overshoots along the coordinates x. To varify beforehand whether an overshoot arises, we should investigate the mutual location of the domain G, coordinate hyperplanes, and the initial point, taking into account the expected character of motions beyond G. In regard to a multilevel binary system, this investigation involves appreciable efforts. We are of the opinion that it is worthwhile to leave aside this issue and assume that any regeneration at $\|s\| > \sigma$ is undesirable. What counts in favor of our opinion is that rotations about G commonly give rise to oscillations of the coordinates x, which is undesirable even if an overshoot is absent — recall the example of an automobile moving from row to row along a highway. Besides, the origin of coordinates in the state space of a system belongs to the domain G, for which reason the rotations about G necessarily give rise to overshoots under certain initial conditions.

Thus, it makes sense to state that oscillations of the coordinates s are undesirable. Do they actually appear? In other words, is it conceivable that the induction algorithm of the form (9.9) exhibits some properties which we have not yet revealed and which rule out the possibility of the emergence of undesirable oscillations?

Unfortunately, miracles do not exist. The algorithm (9.9) allows a system to rotate quite freely about G or perform various oscillations. Indeed, the directional vector of the inducing control (9.9) is $-s/\|s\|$; in the space of the coordinates s, the vector u at any point is directed toward the coordinate origin. A value of the quantity $\|u\|$ is high enough to cause the system to move to the point $s = 0$, i. e., to ensue a decrease in $\|s\|$. But motions of the system in the directions orthogonal to the vector s depend on the properties of the unknown right side of the object model, so that any unforeseen events may occur.

Let us examine this situation in detail. We describe the object of interest by the same equations as those presented in Sec. 9:

$$\dot{x}_A = \varphi_A(x, t)$$
$$\dot{x}_B = \varphi_B(x, t) + Bu$$

The control (9.9) imparts the object a component of the velocity in the direction $-Bs$. This direction is not known exactly because B is indefinite and possibly varies all the time. All we know about the vector Bu is that its projection on the vector s is fairly large.

The system motions orthogonal to s are defined by a sum of two orthogonal projections of the phase velocity vector. The first projection is a vector $\dot{x}_A = \varphi_A$ that is orthogonal not only to the vector s at a certain instant, but also to the entire subspace generated by the coordinates s and, evidintely, by the coordinates x_B and u. There is no way of applying directly a control action to $\dot{x}_A$ so as to exert the desired effect on it. The only way to act on x_A is to transfer the control action to x_A through x_B. Such is indeed the case in the process of generation of internal feedback. But now we consider the motion beyond G when this process does not yet set in. Is it possible to limit somehow or other the arbitrary behavior of the system here too? In the general case, we cannot evidently impose limits. We leave aside a more in-depth analysis of this issue. So, if the system undergoes oscillations along the coordinates x_A, we have to accept the fact.

The second projection $\dot{x}$ orthogonal to s is an element of the space of the coordinates s, i. e., a component of the system motion, which is amenable to a direct action of the control. The control (9.9) exerts an action on the above component when the vector Bu is noncollinear with respect to u. But we do not know and are not able to know the distinctive feature of this action. In this case, the motions cause the system to rotate in the same way as we described at the begining of Sec. 17.2. To preclude the rotation of the system, we should change the control algorithm, namely, abandon the control (9.9).

The plots illustrative of oscillations in induction systems are shown in Fig. 17.2; the plot relating to a termwise induction procedure is shown in Fig. 17.3.

17.3. Ways of Eliminating Oscillations

We will consider the ways of how to exclude oscillations of a system in the space of coordinates s at a high amplitude that is greater than an induction

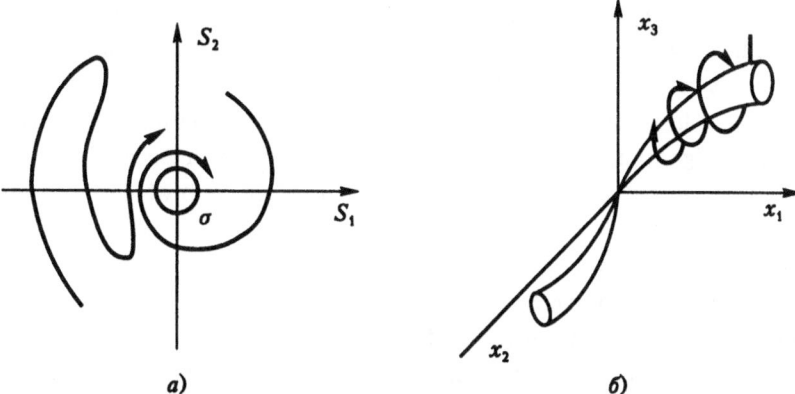

Fig. 17.2. Oscillations (*a*) in induction systems within coordinates arise because the effect of the control on motions of a system in the directions orthogonal to the vector s depends on indefinite disturbances. Rotations (*b*) of the point s about the σ-neighborhood of zero correspond to rotations of the point x about the domain G.

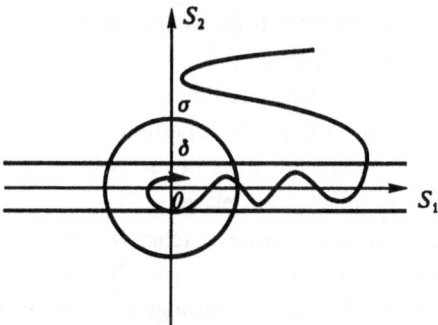

Fig. 17.3. The termwise induction procedure provides a way of precluding the rotation of *s* about the σ-neighborhood of zero and hence the rotation of x about G. The plot illustrates the case where the control causes the coordinate s_2 and then the coordinate s_1 to approach zero. The control fails to eliminate completely oscillations, but ensures that the amplitude of the overshoot along the coordinate s_2 does not exceed the prescribed value $\delta < \sigma$.

error σ. Our objective is to limit the freedom of motions of the point s when it approaches the σ-neighborhood of the coordinate origin.

Let us recall that in Sec. 13 we suggested a similar algorithm when we considered the problem for saving control actions from the outset (see Sec. 13.5). We pointed out in Sec. 13 that the control holds the point within the ball of radius σ, the center of which moves according to the assigned law from the initial point to the coordinate origin within the space of an induction error signal, so that high-amplitude oscillations are actually absent. But, as noted in Sec. 13, the suggested algorithm is very vulnerable to faults. The system is able to rotate about the ball as soon as it falls out of it. Thus, there is reason to confirm our reluctance to employ the algorithm with the "long-term memory", in which the current control depends on the system coordinates at the initial instant from which much time elapses.

We resort to another idea. Note that the motion of coordinates orthogonal to the direction of the vector $s(t)$ is possible only in the systems where this orthogonal direction exists at all. It does exist only in spaces of the dimension that is larger than unity. If s is a one-dimensional vector, then the scalar control of the form (9.9) ensures a monotonic decrease in $\|s\|$, and so no oscillations arise. Passing on to the general case, we replace the vector induction problem by a sequence of a few one-dimensional problems. We can now use, for example, the coordinate u_1 so as to bring the vector s_1 closer to zero and hold it in place, then use u_2 to bring s_2 closer to zero, and so on. As a result, conditions of the form $|s_i| \leqslant \sigma_i$ for all i will prevail in the system; a requisite choise of $\{\sigma_i\}$ will ensure the required condition $\|s\| \leqslant \sigma$.

We will call the described procedure the termwise induction, bearing in mind that it causes the sequential initiation of conditions in the system, which serve as conjunctive terms of the assertion sufficient to certify that the induction of the desired type of internal feedback takes place.

One cannot affirm that the termwise induction eliminates all oscillations in the space of s. Indeed, when the control u_1 shifts its own coordinate s_1 to the appropriate neighborhood of zero, the orthogonal addition to the space of the one-dimensional vector s_1 is defined not only by coordinates x_A but also by coordinates s_2, s_3, etc. Oscillations along these coordinates are as yet possible. We can only warrant that the point s will not rotate about the σ-neighborhood of zero. The thing is that the hyperplane $s_1 = 0$ divides the space into two subspaces and the control u_1 does not let the system move too far away from this hyperplane. Rotations about the point $s = 0$ near the hyperplane $s_1 = 0$ cannot occur, too, because controls such as u_2, u_3, and others are in operation.

Note that there is no need to carry out the procedure of the termwise induction only through the use of a sequence of one-dimensional problems. It is possible to decompose the vector s into a sum of vector projections s^1, s^2, etc., and the vector u into a sum of u^1, u^2, etc., so that each control u^i will cause the requisite vector s^i to approach zero. The number of conjuctive terms of the assertion that must hold here is found to be less than it is in the case considered above. It is well to use this vector procedure if there is additional information on an object, which permits us to ignore the rotation in spaces of vectors s^1, s^2, etc.

17.4. Termwise Induction and a New Class of Problems

The use of the procedure described above makes it possible not only to preclude the rotation of the point s, but also to extend appreciably the class of objects to which the suggested approach is applicable. Assume that the one-dimensional projection u_1 of the control vector must force the coordinate s_1 to approach zero in the course of the termwise induction. To take advantage of the algorithm of the form (9.9), we have to be confident that u_1 enters into the right side of the requisite equation in a linear fashion together with a coefficient whose value exceeds a certain positive number. It does not matter how the right side of the model depends on all of the remaining coordinates of the vector u: in the course of the synthesis of u_1, these coordinates can be regarded as additional state parameters. Other requirements placed on the right side of the model appear because u_2, u_3, and other controls must sequentially solve their one-dimensional induction problems. Despite this fact, the termwise induction procedure proves feasible for an object model in which it is not the mere vector Bu with the diagonal matrix that contributes to the control in the right side of the model; the right side may even be nonlinear in u.

As in the case of the problem for the elimination of oscillations, the procedure of the termwise induction with conjuctive vector terms exhibits lower capabilities. However, this procedure is relatively simple and deserves attention if it is applicable to a particular object.

Thus, the idea of the termwise induction gives grounds to hope that there is a possibility of extending the field of application of the suggested approach to objects with more restrictive requirements. In the next section we will consider the implementation of this idea.

18. Termwise Induction in Systems with Branching Structures

The termwise induction procedure described above is just a speculative procedure and not an actual one at all. We only pointed out that a definite character of operation of a controllable system can offer practical advantages. Is it possible to implement this type of operation? We should develop requisite algorithms and verify whether they display the desired properties. These issues are dealt with in the text given below.

18.1. General Design Considerations

We will clarify the properties that the mapping $(x, t) \mapsto u$ must possess so as to ensure the termwise induction of a certain internal feedback loop in a system.

Assume that the condition $\|s\| \leqslant \sigma(x_A, t)$ must first be set in a closed system, where $s = x_B - v(x_A, t)$ and x_A and x_B are the projections of the vector x. Next, the coordinates of x_A must obey some condition of whatever type for the present. Let us also assume that it is possible to decompose the control vector into a sum of two orthogonal projections u_A and u_B so that the object model can obtain the form

$$\dot{x}_A = \varphi_A(x, u_A, t)$$
$$\dot{x}_B = \varphi(x, u_A, t) + B(t)u_B \qquad (18.1)$$

where $B(t)$ is the diagonal matrix with diagonal elements in the section $[1/b, \ b]$, $b > 0$. We can now resort to the considerations set forth in Sec. 9 and construct the control $u_B = u_B(x, u_A, t)$ that ensures the inequality $\|s\| \leqslant \sigma$ after a time or affords the motion of $x \to 0$ along an exceptional trajectory. For this, we need to have only such a body of information on φ_A and φ_B that enables us to define the majorants $\Phi(x, u_A, t)$ of their norms. The developed control suppresses the effect of u_A on the object in the same way as it does the effect of disturbances; it makes no difference how φ_A and φ_B depend on u_A. Of importance is only the fact that values of these functions are independent of u_B.

Thus, u_B does its job as it suppresses the effect of u_A. How will then u_A handle its task? A similar behavior on the part of u_A will not generally lead to the desired result: the equality $u_A = u_A(x, u_B, t)$ may prove inconsistent with the law $u_B = u_B(x, u_A, t)$. Recall that the control u must interrelate somehow the coordinates x_A only after u secures the inequality $\|s\| \leqslant \sigma$. Substituting the equality $x_B = \mu\sigma + v$ into the first equation of the system (18.1) yields a dynamic object model of the form

$$\dot{x}_A = \varphi(x_A, \mu\sigma + v, u_A, t) \qquad (18.2)$$

If the control $u_A = u_A(x_A, t)$ imparts the solution $x_A(t)$ of equation (18.2) the required properties under any change of $\mu = \mu(t)$ within a unit ball, the stated aim is achieved.

The procedure of developing the control law u_A obviously requires the introduction of constraints on the dependence of values of φ_A on those of u_A. Thus, to implement once again the idea of the induction of some internal feedback, we need to single out a definite projection u_A of the vector u, which, just as the projection u_B of u, enters into the right sides of the equations in a linear fashion together with the diagonal matrix of coefficients. In any case, the dependence of φ_B on u_A is of no significance.

In the developed system, u_A begins to play the specified role only after u_B ensures the fulfillment of the condition $\|s\| \leqslant \sigma$. Until this takes place u_A will have a certain effect on the behavior of the system, but will be unable do anything useful on the whole, nor will it impair the operation of u_B.

18.2. Statement of the Termwise Induction Problem

We will perform the synthesis of a termwise induction system in the general vector case. Assume that an object model can take the form

$$\dot{x}^{00} = \varphi^{00}(x, t)$$
$$\dot{x}^{01} = \varphi^{01}(x, t) + B^1(t)u^1$$
$$\dot{x}^{02} = \varphi^{02}(x, u^1, t) + B^2(t)u^2 \qquad (18.3)$$

$$\cdots\cdots\cdots\cdots\cdots\cdots\cdots\cdots\cdots\cdots\cdots$$

$$\dot{x}^{0m} = \varphi^{0m}(x, u^1, u^2, \ldots, u^{m-1}, t) + B^m(t)u^m$$

where x^{00}, $x^{01}, \ldots, x^{0m}$ are projections of the vector x and $u^1, \ldots, u^m$ are projections of the vector u and $B^1(t)$, $B^2(t), \ldots, B^m(t)$ are diagonal matrices with diagonal elements within the section $[1/b, \, b]$, $b > 0$. The system of equations (18.3) is the most common form of the model for an object among the objects for which the control synthesis is made. When $m = 1$, the system (18.3) reduces to (16.3).

Consider the problem of inducing internal feedback

$$\begin{pmatrix} x^{01} \\ x^{02} \\ \vdots \\ x^{0m} \end{pmatrix} = v(x^{00}, t) \qquad (18.4)$$

at an error of $\sigma(x^{00}, t)$, assuming that v and σ are prescribed functions. The vector on the left side of (18.4) consists of all coordinates of the projections x^{01}, $x^{02}, \ldots, x^{0m}$. It is impossible to construct the control $u = u(x, t)$ of the form (9.9) for the object of interest because the right side of the model depends on the control in the way other than the pertinent one. We will carry out the termwise induction procedure. Let first u^m come into action, then u^{m-1}, etc.

We assign our objective to each projection. For this, we write the vector equality (18.4) as a system of the vector equalities

$$x^{01} = v^{01}(x^{00}, t)$$
$$x^{02} = v^{02}(x^{00}, x^{01}t)$$

$$\cdots\cdots\cdots\cdots\cdots\cdots\cdots\cdots\cdots \qquad (18.5)$$

$$x^{0m} = v^{0m}(x^{00}, x^{01}, \ldots, x^{0m-1}, t)$$

Next, we prescribe functions

$$\sigma^{01}(x^{00},t), \quad \sigma^{02}(x^{00},x^{01},t), \quad \ldots, \quad \sigma^{0m}(x^{00},x^{01},\ldots,x^{0m-1},t)$$

so as to induce internal feedback (18.4) at an error of σ for all $j = 1,2,\ldots,m$ from the conjunction of inequalities $\|s^{0j}\| \leq \sigma^{0j}$, where $s^{0j} = x^{0j} - v^{0j}$. There are many ways of achieving this aim. An infinite number of equalities of the form (18.5) correspond to (18.4). We select an equation that is more suitable for us, for example, on the basis of the geometric forms of trajectories, because a closed system will sequantially move near manifolds such as

$$\|s^{0m}\| = 0, \quad \|s^{0m}\| + \|s^{0m-1}\| = 0, \quad \|s^{0m}\| + \|s^{0m-1}\| + \|s^{0m-2}\| = 0$$

Functions $\{\sigma^{0j}\}$ can be so chosen that a "closer" expression should have a desired meaning. We will not give reasons for the choice of a specific set of the pairs of functions $\{v^{0j}, \sigma^{0j}\}$ and assume that this choice has been made.

18.3. Synthesis and Properties of the Control Algorithm

Relying on the line of reasoning set out in Sec. 9, we build up the control

$$u^m = -\frac{s^{0m}}{\|s^{0m}\|} \widetilde{\Psi}\big(\|s^{0m}\|, \sigma^{0m}\big) F_m(x, u^1, u^2, \ldots, u^{m-1}, t) \qquad (18.6)$$

Note that in determining the function Fm, we should use majorants of the norms of the partial derivatives v^{0m} and σ^{0m} and also the majorant of the quantity $\|\varphi^{0m}\|$ and norms of the right sides of all eqautions (18.3), excepting the last one. Under requisite conditions, the control (18.6) either secures the fulfillment of the condition $\|s^{0m}\| \leq \sigma^{0m}$ in the course of time, regardless of the nature of changes in $u^1, u^2, \ldots, u^{m-1}$, or causes x to go to zero along an exceptional trajectory.

If $\|s^{0m}\| \leq \sigma^{0m}$, then the vectors $x^{00}, x^{01}, \ldots, x^{0m-1}$ vary in the same way as state projections of a certain representative of the family of the dynamic systems

$$\dot{x}^{00} = \varphi^{00}(x^{00},x^{01},\ldots,x^{0m-1},v^{0m}+\mu\sigma^{0m},t)$$
$$\dot{x}^{01} = \varphi^{01}(x^{00},x^{01},\ldots,x^{0m-1},v^{0m}+\mu\sigma^{0m},t)+B^1(tu^1)$$
$$\dot{x}^{02} = \varphi^{02}(x^{00},x^{01},\ldots,x^{0m-1},v^{0m}+\mu\sigma^{0m},u^1,t)+B^2(t)u^2 \qquad (18.7)$$

$$\cdots\cdots\cdots\cdots\cdots\cdots\cdots\cdots\cdots\cdots\cdots\cdots\cdots\cdots\cdots\cdots$$

$$\dot{x}^{0m-1} = \varphi^{0m-1}(x^{00},x^{01},\ldots,x^{0m-1},v^{0m}+\mu\sigma^{0m},u^1,\ldots,u^{m-2},t)+B^{m-1}(t)u^{m-1}$$

for all kinds of $\mu = \mu(t)$, $\|\mu\| \leq 1$. Equations (18.7) result from the first m equations (18.3) by substituting $x^{0m} = v^{0m}+\mu\sigma^{0m}$. The problem of the synthesis of u^{m-1} for the system (18.7) is similar to that of the synthesis of u^m for (18.3). We have

$$u^{m-1} = -\frac{s^{0m-1}}{\|s^{0m-1}\|}\tilde{\psi}\left(\|s^{0m-1}\|, \sigma^{0m-1}\right)F_{m-1}(x^{00}, \ldots, x^{0m-1}, u^1, \ldots, u^{m-2}, t)$$

(18.8)

As the condition $\|s^{0m}\| \leqslant \sigma^{0m}$ sets in, the control u either ensures the inequality $\|s^{0m-1}\| \leqslant \sigma^{0m-1}$ after a time or causes $x \to 0$ along an exceptional trajectory.

The same procedure is used to synthesize u^{m-2}, u^{m-3}, etc.

Inequalities $\|s^{0i}\| \leqslant \sigma^{0i}$ sequentially appear in a closed system for $i = m$, $m - 1$, etc., until they all emerge if some control u^j does not preclude this procedure by forcing the system to move to the coordinate origin along an exceptional trajectory. The area of application of the induction idea extends to a wide class of objects defined by (18.3). The proper subdivision of u into projections permits avoiding the rotations described in the preceding section.

Let us note that the projection u^{m-1} has a useless effect on an object at $\|s^{0m}\| > \sigma^{0m}$. Moreover, this effect can sometimes be considered harmful because the control u^m has to exert this effect in any case, which leads to rather high values of $\||u|^m\|$. In fact, all controls u^j at $j < m$ are useless in this situation, and they persistently suppress one another. In practical cases, this behavior is quite undesirable. It is possible to remedy the situation by introducing into the control law a cofactor — a "switch" — for each projection u^j. In the simplest case, a characteristic function of the domain of the state space can act as the switch. In this domain, a certain projection must be in operation, i. e., a function that is equal to unity in this domain and to zero at all of the remaining points. For a projection u^j there is a need for the characteristic function of the set of points of the form $\bigcap_{i>j}\{x \mid \|s^{0i}\| \leqslant \sigma^{0i}\}$. If the control discontinuities are taken to be inadmissible, then use can be made of cofactors (switches) instead of characteristic functions, the values of which continuously grow from zero to unity when the system approaches an appropriate domain.

18.4. Structure Representation

The language of block diagrams with the use of the binary principle offers a convenient way of describing a termwise induction algorithm. Each control projection u^j is given as the coordinate-feedback operator output

$$u^j = U^j_{\Psi^j}(x, u^1, u^2, \ldots, u^{j-1}, t)$$

(18.9)

with the parameter Ψ^j set up in the coordinate-operator feedback loop (Fig. 18.1, 18.2). The function Ψ^j will be defined as a product of the smoothing directional cofactor and the cofactor, referred to as the switch, if it is available.

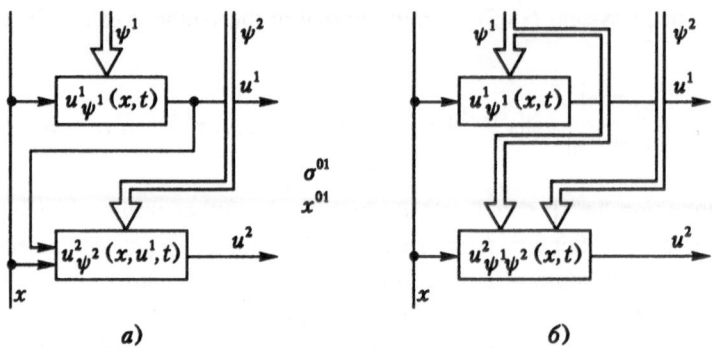

Fig. 18.1. Branching structures serves to carry out the procedure of termwise induction. The figure demonstrates two versions of a block-diagram section of the system: the version (a) involving the branching for the transfer of a control projection and the version (b) involving the branching for the transfer of a projection of the operator signal. Both of the versions are adequate for one and the same algorithm in which the control projections u^1 and u^2 successively induce the requisite projections of internal feedback, where u^2 acts first.

According to (18.9), the output of each of the m coordinate feedback operators, excepting the last one, serves as an input action not only for a controllable object, but also for all of the coordinate feedback operators with high values of the upper indices. Representing graphically these elements, we obtain structures (block diagrams) which can rightfully be spoken of as branching structures. In this case, the branching occurs during the transfer of projections of the control vector.

The set of equalities of the form (18.9) implicity defines the mapping

$$u = U_{\varphi^1 \varphi^2, \dots, \varphi^m (x, t)}$$

that can be converted to an explicit form. For this, we substitute expression (18.9) for $j = 1$ into a similalr expression for $j = 2$, use the substitution result in (18.9) for $j = 3$, and so on. Finally, we obtain the reprsentation of the form

$$u^j = \widetilde{U}^j_{\varphi^1 \varphi^2 \dots \varphi^j (x, t)} \qquad (18.10)$$

The structure will also assume the branching form, but the branching will occur during the transfer of operator signals.

18.5. Termwise Induction in Tracking Systems

In Sec. 16 we considered the ussue of constructing a tracking system out of a sequence of induction algorithms (Fig. 18.3). The approach described in Sec. 16 can be used to construct a system by interconnecting complete blocks

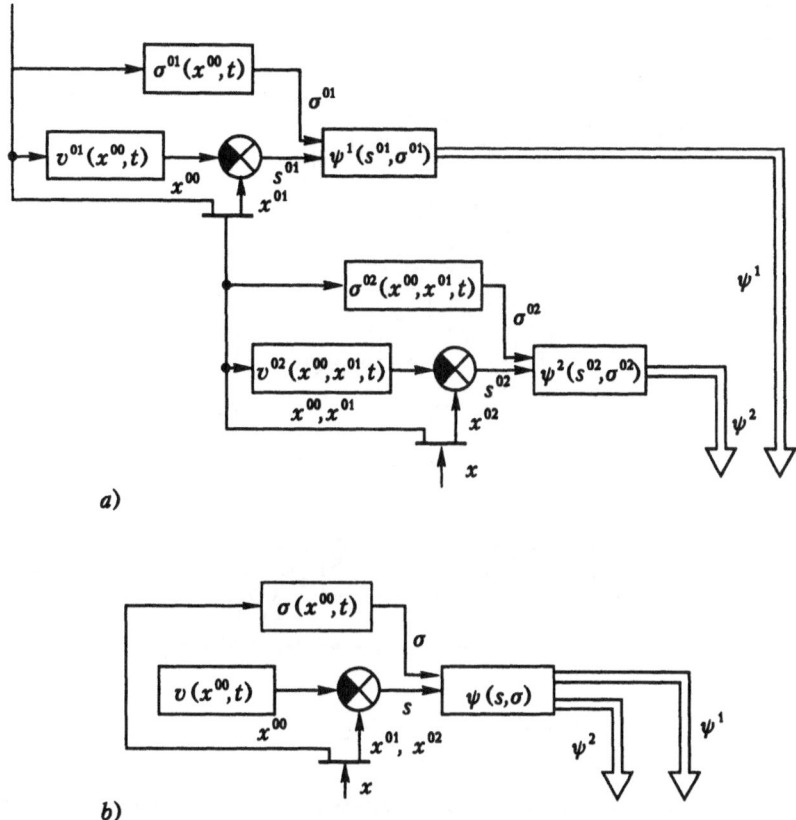

a)

b)

Fig. 18.2. The coordinate-operator feedback loop (a) of a binary system with the branching structure consists of two groups of operators each of which is suitable for the generation of operator signals in a system without the branching of the structure. It is not always expedient to point out all details of the signal shaping in the block diagram. Sometimes, it makes sense to represent the structure in a simplified way, for example, through the replacement of the block diagram (a) by the conventional block diagram (b). The diagram (b) does not account for the actual sequence of transformations of signals.

of induction algorithms rather than individual algorithms intended to perform termwise induction procedures. This approach obviously offers the possibility of extending the class of controllable objects and eliminating some oscillations of an image point. In the system thus constructed, the control and the sequence of control-induced feedback loops ensure, on the whole, the termwise induction. The graphic representation of the interacting elements of such a system takes the form of the block diagram of a structure that can by right be given the name multilevel branching structure.

In the multilevel structure, any internal feedback operator, excepting the operator of the highest level, simultaneously plays two roles: serves as a means

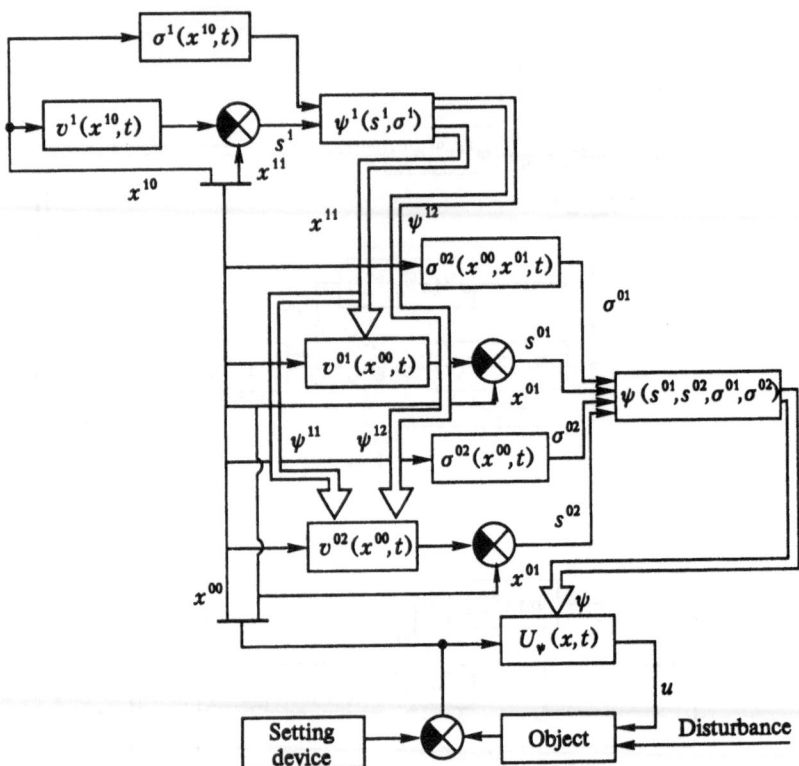

Fig. 18.3. The multilevel branching structure serves to implement solutions of tracking problems. The branching relates to a structure level at which the system carries out the procedure of the termwise induction in the chain of interconnected induction algorithms. The diagram illustrates the case of the structure branching at the first level of the coordinate-operator feedback loop, with the substructure at the upper level displayed conditionally. In addition, this diagram indicates the boundaries of the clarity of structure representations: if a system is somewhat more complex, its block diagram will hinder the perception of the formal description.

of carrying out the procedure of the induction of the next internal feedback loop and as a means that accepts the result of the induction procedure performed by a lower-level operator. If both of the procedures involve the termwise induction, then the operator is broken up twice into orthogonal projections by two different methods. For example, the decomposition of the operator v defined by (18.4) into the projections (18.5) mainly depends on the way of how the coordinates of u enter into the right side of the object model. But if this operator acts as an inducing control, its decomposition into projections depends on the fact of how the vector coordinates x^{01}, $x^{02}, \ldots x^{0m}$ enter into the right side of the equation $\dot{x}^{00} = \varphi^{00}(x, t)$.

During the control process, the internal feedback operator of the $(j + 1)$th structure level is first built up of its own projections v^{jm_j}, v^{jm_j-1}, ..., v^{j1} under the action of the operator of the jth level and then breaks down into projections $v^{jm_{j+1}}$, $v^{jm_{j+1}-1}$, ..., v^{j1} so as to shape up the operator of the next $(j+2)$th level. We use the designations in which the first upper index is less by unity than the ordinal number of the structure level in question and the second upper index is equal to the ordinal number of a vector projection.

Let us note that we actually used the procedure of the termwise induction in a tracking system in Sec. 16, too, for the isolation of "indifferent" components of the inducible internal feedback operator, which do not find use as means for the induction of internal feedback at a successive level.

18.6. Intact Clusters

The multilevel branching structures described above implement systems in which the individual induction algorithms are tied together into clusters. But we described by far not all the systems (clusters). For example, we did not consider the cluster systems in which the projections of operators at different structure levels jointly perform desired functions (induce a certain feedback loop). But let us recall what we poited to in Sec. 17: there are too many cluster systems and it makes no sense to examine theoretically their general features. We hope that our lecture course will be of service to those who will attempt to construct rather complex cluster systems for practical objects, including systems that we did not describe in detail.

Induction Control.
Practical Examples

In the last chapter of our lecture course, we describe various methods for disclosing the feasibility of the theory outlined above. Our objective is to examine these methods and substantiate the ways that we have chosen.

In regard to actual practice, it seems to us that the use of the described mathematical tools for the problems of control of very complex multidimensional dynamic systems can afford the most benefit. But we do not examine these systems because it would be necessary to adapt to a specific situation the fairly complex structures considered in Secs. 16 and 18. It would be necessary to consider a mathematical model, conditions of its action, and other aspects. The description of each example of a multidimensional system would take up tens of pages of the text that would be far from simple to the perception. It can be said with reasonable confidence that only specialists who engage in the subject of systems would want to read such a text.

For this reason, we will consider examples relating only to the idea of internal feedback induction rather than to the entire range of the mathematical tools outlines in the book. This permits us to restrict the discussion to models of the second or the third order. Of course, simple models cannot afford a fairly complete description of not in the least simple reality. Therefore, we will take examples of the elements more similar to toys rather than of the elements encountered in actual practice.

The use of simple examples bring definite advantages. They enable us to gain a knowledge from the study of real objects (even if toy-like ones), the construction of mathematical models, and the solution of control problems. The properties of simple systems are "visible" to the unaided eye, as are the properties of the control laws derived on the basis of the theory outlined above.

Thus, the subject set out in this chapter does not require a special training for its understanding, and so there is reason to hope that this subject will attract attention of many readers.

19. Biology:
Control of Predator-Victim Systems

We examine a two-component biocenosis, i. e., a dynamic system descrbing the interaction of two populations of different species. Our interest lies in time variations of the sizes of these populations. In this lecture, we use the simplest Volterra model.

19.1. Discussion of Practical Issues

The interaction of two populations of animals represents one of the simplest examples of a dynamic system in biology, but the investigation of such a system holds a certain interest for practical problems. The control problems treated in this section relate to the practical problems involved with the biological protection of plants and the stabilization of quantities of wild animals at the zero level. We will first describe the eesence of these problems in a comprehensible verbal form.

A diversity of herbivorous insects, from caterpillars to colusts, invade cultivated plants and forests. A lasting dream of a farmer and a forester is to reduce the populations of insect pests. How is it possible to realize this dream? A widespread way is to "combat" insects with chemicals. However,, this way of pest control is not always acceptable. First, while exterminating pests, the pest control chemicals adversely affect the health of a human being. Second, insects adapt to insecticides and their populations rapidly regain their size even in the course of their extermination: initially scanty mutants immune to poisons begin to multiply rapidly and restore the population size. That is why more and more attention is given to other plant protection methods of which one method has its origin in the use of insects that are natural enemies of insect pests. The essense of this method lies in the artificial stimulation of the growth of insects — beasts or parasites — which poison the life of herbivorous pests and hence cut down their populations. These useful insect beasts or parasites, which will be called "predators" in the text below, are bred in special incubators and released from the incubators to the environment in large quantities. The number of the predators heavily increases and becomes well above the natural level, so that they effectively reduce the population of their "victims", i. e., insect pests. It is useful to clarify how many incubation predators should be set free per unit time so as to stabilize the number of victims and keep it near the zero level. This question can be given the shape of a control problem.

A feature specific to the preditors under consideration is that we can release to the environment the desired number of the preditors, but cannot remove any definite number of these insects from their biocenosis. But if we have to do with large animals, the situation often becomes the opposite of the above one: the widespread method of exerting actions on a community of these animals is to trap or shoot a hunting quata. Is it possible to stabilize the sizes of two interacting populations by resorting to these ways of the removal? As regards two populations of the predator-victim system, the above two ways are applicable either to both the populations of animals, for example of the wolf-deer system, or only to one of these populations. In the latter case, these are only predators, for example, of the fox-mouse system (it is a fox that can prey on mice in a forest) or only victims, for example, of the tiger-boar system (hunting of tigers is forbidden). These practical problems can entail some problems of automatic control.

The change-over from the verbal description to problems always calls for the stage of formalization, i. e., the construction of mathematical models for the systems under study. The procedure involved is dealt with in Sec. 19.2.

19.2. Model of the Predator-Victim Biocenosis

The model outlined below defines the way of how the total numbers N_1 and N_2 of two interacting populations of animals vary with time. Issues such as the settlement of populations and their classification in age are left out of consideration. The state of a system is taken to be the pair of the numbers N_1 and N_2. The system evolution is given by two first-order ordinary differention equations solvable with respect to the derivatives of state coordinates. Here a clear distinction between the model and reality is evident: the exact size of a population in nature is a time function taking only nonnegative integers, which is not amenable to the differentiation. The model includes continuous functions that approximately specify the size of a population.

Let us use the the Volterra equations describing a predator-victim system without disturbances:

$$\dot{N}_1 = N_1(\varepsilon_1 - \gamma_{12}N_2)$$
$$\dot{N}_2 = N_2(-\varepsilon_2 + \gamma_{21}N_1)$$
(19.1)

where N_1 and N_2 are the numbers of victims and predators, respectively, and ε_1, ε_2, γ_{12}, and γ_{21} are positive constants. The system of equations (19.1) displays plausible properties: in the absence of predators, the rate of growth of victims increases exponentially, slows down when the number $N_2 \neq 0$ of predators is low and comes to a halt when $N_2 = \varepsilon_1/\gamma_{12}$. Next, the extinction of the victims takes place when the number of predators grows, but the predactors themselves begin to die out without victims, and their at a larger number of victims. $N_1 = \varepsilon_2/\gamma_{21}$ A phase portrait of the system in a qualitative form appears in Fig. 19.1 where singular points i. e., saddle points and centers, have the coordinates $(0,0)$ and $(\varepsilon_2/\gamma_{21}, \varepsilon_1/\gamma_{12})$, respectively.

It should be noted that some properties of equations (19.1) contradict actual conditions. First, contrary to the assertions that follow from the equations, the

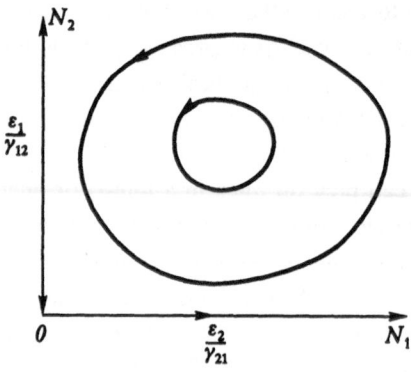

Fig. 19.1. The behavior of a two-component biocenosis in coordinates of the number (N_1) of victims and the number (N_2) of predators. Variations in the size of populations are specific to the system of this type. The typical periods of the variations are a few years for mammals and a few days for protozoa.

growth of each population is subject to the restraint since the animals compete with one another. The competition increases with the size of the population and is the consequence of the shortage of food resources and territories. Second, it is doubtful that the fixed number $N_2 = \varepsilon_1/\gamma_{12}$ of predators is able to stabilize any arbitrary high number of victims, as follows from the first equation; the second equation of the system (19.1) casts similar doubts. However, we will accept equations (19.1) in this lecture. It can be assumed that the problems at hand involve small numbers N_1 and N_2 and the desparity between the model and reality is insignificant. For the model to be more plausible, we suppose that ε_1, ε_2, γ_{12}, and γ_{21} are varying and nonmeasurable parameters of which we only know that they change within definite boundaries.

We will now examine how to account for the effect of a human being on a biocenosis by means of equations. As in the examples considered in Sec. 19.1, the rate of an artificial increase or a decrease in the size of any population can serve as a quantitative characteristic of this effect. This type of control depends on the additional summands u_1 and u_2 in the right sides of the equations

$$\dot{N_1} = N_1(\varepsilon_1 - \gamma_{12}N_2) + u_1$$
$$\dot{N_2} = N_2(-\varepsilon_2 - \gamma_{21}N_1) + u_2 \qquad (19.2)$$

Different practical situation involve different constraints imposed on u_1 and u_1. For example, the constraint $u_2 \geqslant 0$, $u_1 = 0$ corresponds to the reproduction of insect predators, $u_2 \leqslant 0$, $u_1 = 0$ to the shooting of foxes, $u_1 \leqslant 0$, $u_2 = 0$ to the shooting of boars, and $u_1 \leqslant 0$ and $u_2 \leqslant 0$ to the shooting of wolves and the deer. It is also possible to define other constraints when considering certain plausible examples and hence to substantiate the practical meaning of control problems with these constraints. But we will not handle this task and give the ready the oppotunity to take up the question involved on his own.

Thus, we have the model of the system (19.2), the constraints ($N_1 \geqslant 0$ and $N_2 \geqslant 0$) on state coordinates, and the four versions of constraints on control coordinates. It is now possible to discuss and solve automatic control problems.

In the subsequent sections, we will group these problems together according to the features of constraints imposed on a control. In considering these problems, it proves convenient to describe systems in the variables N_1 and N_2, without using equations for the coordinates of a control error signal.

19.3. Reproduction of Predators for Annihilation of Victims

Assume that it is possible to supplement the biocenosis per unit time by u predators raised under the artificial conditions. In this case, the system conforms to the equation

$$\dot{N}_1 = N_1(\varepsilon_1 - \gamma_{12}N_2)$$
$$\dot{N}_2 = N_2(-\varepsilon_2 + \gamma_{21}N_1) + u \qquad (19.3)$$

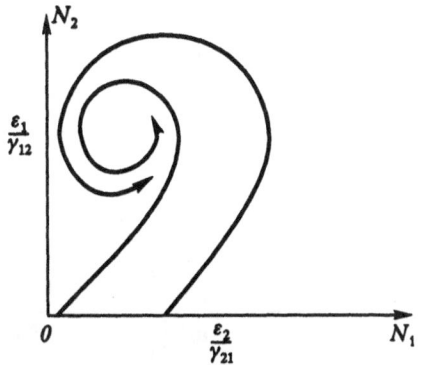

Fig. 19.2. The proportional control conforms to "common sense": the rate of reproduction of predators is proportional to the number of victims subject to extermination. The desired result is not reached at any value of the proportionality factor.

The control problem is to determine the law of variations in $u \geqslant 0$, which would ensure a definite character of changes in N_1 and N_2 in the system (19.3). In particular, the problem of the protection of plants against victims calls for the assurance that N_1 drops to zero; changes in N_2 are not fixed in this case. Let us consider the possibilities of solving this problem.

First of all, we estimate the effectiveness of the proportional control law $u = kN_1$, $k > 0$. This law complies with "common sense": if there are many harrmful victims, it is necessary to raise many predators so that they can prey on victims, but if the number of victims is small, there is nothing else to do, but wait. But this reasoning is not suitable here. At constant parameters $\varepsilon_1, \varepsilon_2, \gamma_{12}\gamma_{21}$, the proportional law causes the system to move at $N_1 > 0$ from all initial points to a stable focus (or sometimes to a node) with coordinates

$$N_1 = \frac{\varepsilon_1, \varepsilon_2}{\varepsilon_1\gamma_{21} + k\gamma_{12}} \quad \text{and} \quad N_2 = \frac{\varepsilon_1}{\gamma_{12}}$$

(see Fig. 19.2). We give the reader the opportunity to perform requisite calculations on his own. Clearly, $N_1 \to 0$ cannot approach zero at any finite values of k. Here, the proportional control employed and proved for many dozens of years does not work.

We turn to the common idea described in the lecture course. Let us verify what motions of N_2 force N_1 vary in the desired manner and then find the control u that ensures these motions of N_2. The first equation of (19.3) obviously describes the dynamics of N_1 as a function of N_2, and so $\dot{N}_1 < 0$ when $N_2 > \varepsilon_1/\gamma_{12}$. Therefore, it is sufficient to subject the motions of N_2 to the condition $N_2 = L = \text{const}$, where $L > \sup \varepsilon_1/\gamma_{12}$ (recall that $\varepsilon_1/\gamma_{12}$ can be variable). As follows from Sec. 8, the equality $N_2 = L$ defines internal feedback induced by the operator that degenerates into the constant L.

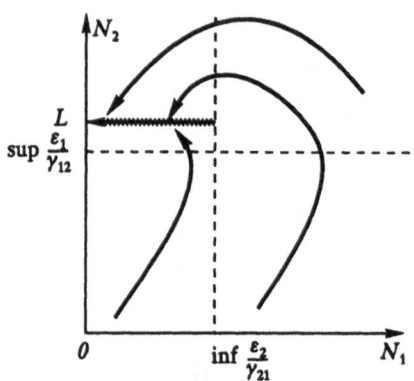

This type of internal feedback appears in the system in due course after the begining of the control process if we ensure that $\dot{N}_2 \geqslant \text{const} > 0$ when $N_2 \leqslant L$. For this, as follows from the second equation of (19.3), it is sufficient to use the control $u = kL$, where $k > \sup \varepsilon_2$. Here, N_2 reaches the required level in a finite time. The further artificial reproduction of predators is useless, and it is expedient to provide a control law with a switch (a cofactor). Let

Fig. 19.3. The system of a varying structure causes the number of victims to drop to zero. But within the segment $N_2 = L$ of the straight line the system functions in the sliding mode and the control is subject to oscillations at an infinite frequency. This system is unrealizable in practice.

$$u = \begin{cases} kL & \text{for} \quad N_2 \leqslant L \\ 0 & \text{for} \quad N_2 > L \end{cases} \qquad (19.4)$$

This control system belongs to the class of varying structure systems (VSS).

Figure 19.3 illustrates a qualitative character of trajectories of the system (19.3), (19.4). As is seen, the image point reaches the segment

$$N_2 = L, \quad 0 \leqslant N_1 \leqslant \varepsilon_2/\gamma_{21}$$

in a finite time and shifts to its left end. In this case,

$$N_1(t) = O(\exp(\sup \varepsilon_1 - L \inf \gamma_{12})t) \qquad (19.5)$$

The basic disadvantage of the control law (19.4) is the discontinuity of the control action in the segment $N_2 = L$ of the straight line; the system moves along this segment in a sliding way. As a result, an enterprise for the reproduction of predators switches over from the complete standstill to the reproduction of kL animals in a unit time, and vice versa, in which case the switching frequency must be infinite. Clearly, no enterprise can operate in this manner, for which reason the control law (19.4) has no practical value.

To change the situation, we introduce a range of transient control values, which makes it possible to convert from the standstill to the work on the reproduction of predators. Let this range represent a band $|N_1 - L| \leqslant \sigma$, $\sigma > 0$, $L - \sigma > \sup \varepsilon_1/\gamma_{12}$, and the control law have the form

$$u = kL \max\left\{0, \min\left\{1, \frac{\sigma + L - N_2}{2\sigma}\right\}\right\} \qquad (19.6)$$

The system devoid of discontinuities behaves so as displayed in Fig. 19.4. In a finite time, the image point penetrates into the rectangle $L - \sigma \leqslant N_2 \leqslant L + \sigma$, $0 \leqslant N_1 \leqslant \varepsilon_2/\gamma_{21}$, in which it moves to the left. The estimate (19.5) now obtains the form

$$N_1(t) = O\big(\exp(\sup \varepsilon_1 - \inf \gamma_{12}(L - \sigma))t\big)$$

The sequence of measures taken to reproduce predators, i. e., to implement the control (19.6) at a fairly large initial number of victims and a small number of predators is as follows: first, the number of predators is made to grow to kL in a unit time (portion AB of the curve), then the reproduction slowly decreases to zero (portion BC), remains at the zero level for some time (portion CD), grows again, and later on its level varies between 0 and kL (below the point D). It is of importance for practical purposes that the number of predators should be held at a level of $N_2 \geqslant L - \sigma$ even if almost no victims

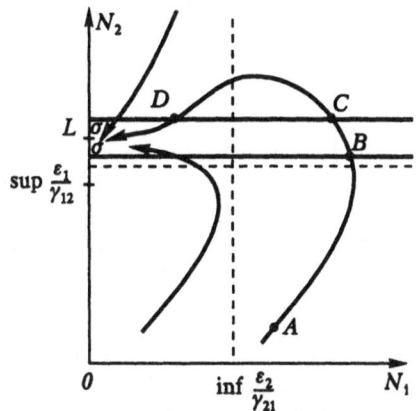

Fig. 19.4. The binary system with a continuous control is devoid of drawbacks of a varying structure system.

are left in nature. In this case, an appreciable number of predators are doomed to die out of starvation. This action is certainly contrary to humanity. However, the objective by itself for the complete annihilation of victims is also vulnerable from the ethic viewpoint. One can seek consolation in the fact that the use of insecticides certainly leads to still much worse aftereffects. A complete solution of these ethic problems and practical ones, too, evidently depends on the basic changes in the technology of plant growing. For now, we have to choose the lesser of many evils.

19.4. Where is it Possible to Stabilize Predator-Victim Systems?

A complete extermination of victims is not always desirable even if they are agricultural pests. A certain number of insects need be preserved in the genofond of nature. Before clarifying how to preseve them, we should estimate our possibilities, i. e., understand what can be done by way of certain effects on a biocenosis. We will look at this issue and discuss the objectives that are worth envisaging in control problems.

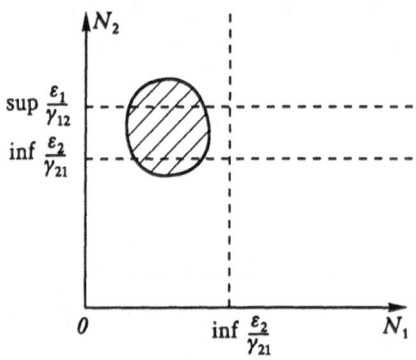

Fig. 19.5. Freedom of the system motion can be bounded only within a domain where there are points that can move along either of the two coordinate axes. The plot displays such a domain for systems that act on a population of predators. Since the direct action on victims is absent, there is a need to extend the domain along the axis N_2. A limit on the control (invariance in sign) calls for the extension of the domain along the axis N_1.

We assume that in the course of the control it is necessary to move the image point of the system to the assigned domain of the space of states and hold it there later on. What domains can we assign? Clearly, if $\dot{N}_1 > 0$ everywhere in the closure of a specified domain, the system will not be able to remain in the domain and will be moved away to the right for certain. It is easy to understand that a domain in which the system can be held in place, i. e., the domain imparted the invariance properties, must fit the following requirement: given a certain control law, each phase velocity component (i. e., both $\dot{N}_1$ and $\dot{N}_2$) does not have the same sign at all points of the closure of the domain. As for practical problems, we can disregard the cases where $\dot{N}_1$ or $\dot{N}_2$ reduces to zero at the domain boundary and consider a simpler case: $\dot{N}_1$ and $\dot{N}_2$ in the domain are sign-variable. This condition is certainly insufficient, but it enables us to evaluate the possibility of stabilizing the predator-victim system.

For example, the number $\dot{N}_1$ is obviously independent of the control in the system intended to reproduce victims. Therefore, the assigned domain must contain a segment $N_2 = \varepsilon_1/\gamma_{12}$ of the line in which values of $\dot{N}_1$ vary in sign. Because values of the system parameters are changed, but not measured, the location of this line is unknown, and so this domain should contain a section of the band whose width corresponds to the set of values of the fraction $\varepsilon_1/\gamma_{12}$. In the above system, the inequality $\dot{N}_2 > 0$ is valid for any $u > 0$ if $N_1\varepsilon_2/\gamma_{21}$. Consequantly, the above section of the band must lie to the left of $N_1 = \inf \varepsilon_2/\gamma_{21}$, as dipicted in Fig. 19.5.

Similar considerations are suitable for verifying limits on the choice of objectives and for solving other control problems.

19.5. Stabilization of Systems Involving Reproduction of Predators

Suppose that we have to stabilize in the system (19.3) a number of victims at a level of $N_1 = M$ accurate to δ, i. e., to ensure that the condition, $|N_1 - M| \leqslant \delta$ holds after a certain time. Suppose also that $M + \delta < \inf \varepsilon_2/\gamma_{21}$ is valid. Let us determine the operator $v(N_1)$ of internal feedback $N_2 = v(N_1)$ the generation of which at an error $\sigma = \text{const} > 0$, i. e. the fulfillment of an inequality $|N_2 - v(N_1)| \leqslant \sigma$, would afford a desired result. Let constants L_1, L_2, and σ be such that

$$L_1 + \sigma < \inf \frac{\varepsilon_1}{\gamma_{12}}, \quad L_2 - \sigma > \sup \frac{\varepsilon_1}{\gamma_{12}}$$

We set

$$v(N_1) = L_1 + (L_2 - L_1)\max\left\{0, \min\left\{1, \frac{N_1 - M + \delta}{2\delta}\right\}\right\} \qquad (19.6)$$

According to the first equation of (19.3), the fulfillment of the inequality $|N_2 - v| \leqslant \sigma$, which corresponds to the broken band in Fig. 19.6, enables us to find a system under the initial conditions $N_1 \neq 0$ in the domain $|N_1 - M| \leqslant \delta$, $L - \sigma \leqslant N_2 \leqslant L + \sigma$, after a lapse of some time; more exactly, in the slanting section of the band.

It now remains to solve only an induction problem, i. e., to deduce the control that causes the system to reach into the above band and remain there. However, this objective is not always attainable. The thing is that a control action is nonnegative, and so it can push the image point to the band in the direction of growth of N_2 or from the bottom upward, but the point must move on its own in the downward direction. According to the second equation of (19.3), we have $\dot{N}_2 < 0$ when $N_1 < \inf \varepsilon_2/\gamma_{21}$ if $u = 0$. Therefore, the control $u = 0$ enables the system to approach the band at $N_2 - v(N_1) \geqslant \sigma$ and penetrate inside through horizontal portions of its boundary. The situation is not so simple as regards the sloping section: the system does not always move inside from the band boundary. The condition of the required motion is evidently the inequality

$$\left|\frac{dN_2}{dN_1}\right| > \left|\frac{dv}{dN_1}\right|.$$

For this inequality to be valid at the band boundary, it is sufficient that the following assertion should exist:

$$\frac{(L_1 + \sigma)\inf \gamma_{21}\left(\inf \dfrac{\varepsilon_2}{\gamma_{21}} - M - \delta\right)}{(M + \sigma)\sup \gamma_{12}\left(L_2 + \sigma - \sup \dfrac{\varepsilon_1}{\varepsilon_{12}}\right)} > \frac{L_2 - L_1}{2\delta} \tag{19.7}$$

Inequality (19.7) limits both the choice of control law parameters L_1, L_2, and σ and parameters M and δ of the statement of the control problem. All these parameters must define "not too large" a slope of the section of the band into which we want to transfer the image point of the system.

Assume that inequality (19.7) holds. Then, at $N_2 \geqslant v(N_1) + \sigma$, the system with the control $u = 0$ behaves in the desired manner. Let us determine u at other values of N_1 and N_2. In this case, the quantity $s = N_2 - v(N_1)$ must steadily increase in magnitude at $N_2 \leqslant v(N_1) - \sigma$. If $|N_1 - M| > \delta$, then

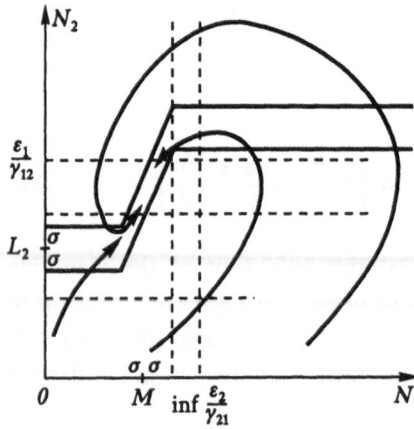

$$\dot{s} = \dot{N}_2 = -N_2\varepsilon_2 + N_1 N_2 \gamma_{21} + u,$$

and at $|N_1 - M| < \delta$, we have

$$\dot{s} = \dot{N}_2 - \frac{dv}{dN_1}$$

$$= -\varepsilon_2 N_2 + \gamma_{21}N_1 N_2 -$$

$$- \frac{L_2 - L_1}{2\delta}\varepsilon_1 N_1 +$$

$$+ \frac{L_2 - L_1}{2\delta}\gamma_{12}N_1 N_2 + u$$

Fig. 19.6.

Besides, $N_2 < L_2$ when $s \leqslant -\sigma$. For this reason, to ensure an increase in s, it is sufficient to set

$$u = U = \text{const} > \varepsilon_2 L_2 + \frac{L_2 - L_1}{2\delta}\varepsilon_1(M + \delta)$$

when $s \leqslant -\sigma$. We assign the control within the band intermediate values in the range between 0 and v and finally derive the law of the form

$$u = U \max\left\{0, \min\left\{1, \frac{\sigma - s}{2\delta}\right\}\right\} \tag{19.8}$$

In general, it would be possible to reduce the control action to zero at $N_1 > \sup \varepsilon_2/\gamma_{21}$ because the quantity s increases in magnitude all the same, but we will not add any cofactors (switches) to the control law so as not to complicate it.

Figure 19.6 illustrates a qualitative character of the trajectories of a closed system. It is seen that the image point can pass once through the band $|s| \leqslant \sigma$. But the system will return to it after a time so as to remain in the band for ever, and then the system will limit its motions within the inclined section of the band. In this case, the control objective will be reached.

19.6. Stabilization of Systems Involving Removal of Predators

Let us look at the model of the dynamic system

$$
\begin{aligned}
\dot{N_1} &= N_1(\varepsilon_1 - \gamma_{12}N_2) \\
\dot{N_2} &= N_2(-\varepsilon_2 + \gamma_{21}N_1) - u
\end{aligned}
\tag{19.9}
$$

where $u > 0$. This system differs from the system (19.3) only in sign during control. How can we stabilize N_1 and N_2 near certain specified values? From the discussion in Sec. 19.4 it follows that the number of victims can be held near a level of $M > \sup \varepsilon_2/\gamma_{21}$, and the number of predators can be held within a section covering the domain of variations in $\varepsilon_1/\gamma_{12}$.

Suppose that there is a need to secure the inequality $|N_1 - M| \leqslant \delta$. We introduce constants L_1, L_2, and σ in the same way as we did in Sec. 19.5. In this case, $L_1 + \sigma < \inf \varepsilon_1/\gamma_{12}$ and $L_2 - \sigma > \sup \varepsilon_1/\gamma_{12}$. Let us determine the function $v(N_1)$ from formula (19.6) and examine the broken band $|s| \leqslant \sigma$ where $s = N_2 - v(N_1)$ (Fig. 19.7). If the system is to be held within the band, then its motions will be bounded within the inclined section in due course. The only difference between this case and the one considered in Sec. 19.6 is

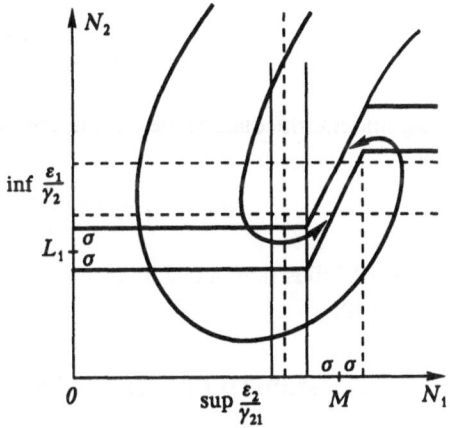

Fig. 19.7.

that here the break of the band corresponds to a higher number of victims and the control forces the system to move in the direction of a decrease in N_2, i.e., from the top downward.

The system must move from bottom to top on its own under the zero control. This situation, as noted above, leads to constraints imposed on the slope of the

band section. In this case, the constraint on the slope takes the form of the
inequality

$$\frac{(L_1 - \sigma) \inf \gamma_{21} \left(M - \delta - \sup \frac{\varepsilon_2}{\gamma_{21}} \right)}{(M + \delta) \sup \gamma_{12} \left(\sup \frac{\varepsilon_1}{\gamma_{12}} - L_1 + \sigma \right)} > \frac{L_2 - L_1}{2\delta}$$

As in Sec. 19.6, this constraint limits arbitrary values of the prescribed parameters M, δ, L_1, L_2 and σ.

When $s \geqslant \sigma$, the system must move so that s should steadily decrease. If $|N_1 - M| > \delta$, then

$$\dot{s} = \dot{N}_2 = -N_2 \varepsilon_2 + \gamma_{21} N_1 N_2 - u,$$

and if $|N_1 - M| < \delta$, we have

$$\dot{s} = \dot{N}_2 - \frac{dv}{dN_1} \dot{N}_1$$

$$= -N_2 \varepsilon_2 + N_2 N_1 \gamma_{21} - \frac{L_2 - L_1}{2\delta} N_1 \varepsilon_1 + \frac{L_2 - L_1}{2\delta} N_1 N_2 \gamma_{12} - u$$

Hence, s decreases when $u = k N_1 N_2$ and

$$k > \gamma_{21} + \frac{L_2 - L_1}{2\delta} \gamma_{12}$$

Assigning intermediate values to the control in the band $|s| \leqslant \sigma$, we derive the law

$$u = k N_1 N_2 \max \left\{ 0, \min \left\{ 1, \frac{s + \sigma}{2\delta} \right\} \right\} \tag{19.10}$$

Figure 19.7 illustrates a qualitative character of the trajectories specified by (19.9) and (19.10).

19.7. Stabilization of Systems Involving Removal of Victims

The system is given by the equations

$$\dot{N}_1 = N_1(\varepsilon_1 - \gamma_{12} N_2) - u$$
$$\dot{N}_2 = N_2(-\varepsilon_2 + \gamma_{21} N_1) \tag{19.11}$$

under the constraint $u \geqslant 0$. As shown in Sec. 19.4, a biocenosis is not subject to the stabilization if the condition $N_2 < \varepsilon_1/\gamma_{12}$ breaks down or the required range of values of N does not include a value of $\varepsilon_2/\gamma_{21}$. In this connection, we consider a problem for which it is necessary to ensure the condition $|N_2 - M| \leqslant \delta$, where $M + \delta < \inf \varepsilon_1/\gamma_{12}$.

In the system (19.11), as distinct from the preceding two systems, the band for the desired motions must stretch vertically. Let L_1, L_2, and σ be positive constants such that

$$L_1 + \sigma < \inf \frac{\varepsilon_2}{\gamma_{21}}, \qquad L_2 - \sigma > \sup \frac{\varepsilon_2}{\gamma_{21}};$$

$$v(N_2) = L_2 + (L_1 - L_2) \max \left\{ 0, \min \left\{ 1, \frac{N_2 + \delta - M}{2\delta} \right\} \right\}$$

As displayed in Fig. 19.8, the residence of the system in the band $|s| \leqslant \sigma$ will force it to obey the desired condition in a finite time, where $s = N_1 - v(N_2)$. The requirement for an increase in s at $s \leqslant -\sigma$ and under the control $u = 0$ imposes the following constraint on the parameters:

$$\frac{(L_1 - \sigma) \inf \gamma_{12} \left(\inf \dfrac{\varepsilon_1}{\gamma_{12}} - M - \delta \right)}{(M + \delta) \sup \gamma_{21} \left(\sup \dfrac{\varepsilon_2}{\gamma_{21}} - L_1 + \sigma \right)} > \frac{L_2 - L_1}{2\delta}$$

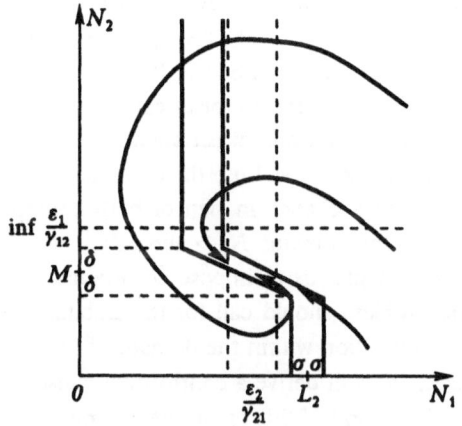

Fig. 19.8.

We agree that this inequality exists and determine a control at which s steadily decreases when $s \geqslant \sigma$. If $|N_2 - M| > \delta$, then

$$\dot{s} = \dot{N}_1 = N_1 \varepsilon_1 - \gamma_{12} N_1 N_1 - u,$$

and if $|N_2 - M| < \delta$, we obtain

$$\dot{s} = \dot{N}_1 - \frac{dv}{dN_2} N_2 = N_1 \varepsilon_1 - \gamma_{12} N_1 N_2 + \frac{L_1 - L_2}{2\delta} \varepsilon_2 N_2 + \frac{L_2 - L_1}{2\delta} \gamma_{21} N_1 N_2 - u$$

Therefore, the control $u = kN_1$ will be adequate, where

$$k > \varepsilon_1 + \frac{L_2 - L_1}{2\delta} \gamma_{21}(M + \delta).$$

Predeterminining u by means of intermediate values in the band $|s| \leqslant \sigma$ yields the control law

$$u = kN_1 \max\left\{0, \min\left\{1, \frac{s + \sigma}{2\sigma}\right\}\right\} \tag{19.12}$$

The plot of Fig. 19.8 demonstrates the behavior of the system (19.11), (19.12).

19.8. Stabilization of Systems
Involving Removal of Predators and Victims

Consider the model

$$
\begin{aligned}
\dot{N}_1 &= N_1(\varepsilon_1 - \gamma_{12}N_2) - u_1 \\
\dot{N}_2 &= N_2(-\varepsilon_2 + \gamma_{21}N_1) - u_2
\end{aligned}
\tag{19.13}
$$

In its form, the system of these equations resembles the elementary system presented in Sec. 5. But the constraints indicating that it is impossible to introduce new animals into a biocenosis markedly distinguish the problem of the object control (19.13) from that considered in Sec. 5. If in Sec. 5 we were able to define an arbotrary direction of the control vector at each point, now we can examine this type of direction only within one direct angle. Therefore, we are unable to apply the above results, and so we will use the contents of Secs. 19.1–19.7.

It is possible to ensure a change in sign of both components of the phase velocity of the system in the domains $N_1 > \sup \varepsilon_2/\gamma_{21}$ and $N_2 < \inf \varepsilon_1/\gamma_{12}$. As stated in Sec. 19.4, this enables us to impose constraints on the statement of a control problem: the problem should call for the stabilization of the system in the vicinity of a certain position within the domain of interest. Let (M_1, M_2) be a point in this domain. We will derive a control that causes the system to move to the prescribed neighborhood of this point after a time.

A natural desire of a researcher is to reduce a new problem to the solved one. For example, it is possible to "expend" the action u_1 on something useful and hence obtain the system for removal of only predators. What are we able to do in this system? We know how to stabilize the coordinate N near the level $M_1 > \sup \varepsilon_2/\gamma_{21}$ (which is what is needed) and the coordinate N_1 within the section containing the set of values of $\varepsilon_1/\gamma_{12}$ (which is not necessary). Let us try to expend u_1 so as to remedy the situation involved with N_2. Recall that the values of this coordinate can be enclosed in the interval containing points at which $\dot{N}_2 > 0$ and points at which $\dot{N}_2 < 0$. Using u_1, it is possible to transfer the domain over the state space, in which $\dot{N}_2$ changes sign. For example, let $u_1 = aN_1$, where $a > 0$. The equations of the system now obtain the form

$$\dot{N}_1 = N_1(\varepsilon_1 - a) - \gamma_{12}N_2))$$
$$\dot{N}_2 = N_2(-\varepsilon_2 + \gamma_{21}N_1) - u_2$$

On deriving the control law u_2, as is done in Sec. 19.6, we can stabilize N_1 near M_1 and limit the motion of N_2 within a section containing the set

$$\left[\inf \frac{\varepsilon_1 - a}{\gamma_{12}}, \ \sup \frac{\varepsilon_1 - a}{\gamma_{12}}\right]$$

Selecting values of the parameter a in the requisite manner, we can transfer the section in which N_2 varies into the neighborhood of the point M_2. The objective thus stated will be attained.

A drawback of the obtained result is evident: it is impossible to reach the stabilization accuracy as high as desired. We examined a similar effect in Sec. 19.5: in the systems that act only on one population of animals, the attainable accuracy depends on the ranges of variations in the model parameters. Thus, it is generally possible to use the method of reducing the problem to the solved one, but its result does not always prove satisfactory.

In this connection, we will resort to intermediate results rather than to those presented in Secs. 19.5–19.7. In all the systems examined above, the control served to solve the induction problem, i.e., forced a system to move along a certain band of width σ, where $\sigma > 0$ is an arbitrarily small number. Forming two such mutually perpendicular bands in the system (19.3), we limit the system motion within the intersection of the bands. Let

$$M_1 - \sigma > \sup \frac{\varepsilon_2}{\gamma_{21}}, \quad M_2 + \sigma < \inf \frac{\varepsilon_1}{\gamma_{12}};$$

$$u_1 = k_1 N_1 \max\left\{0, \min\left\{1, \frac{N_1 - M_1 + \sigma}{2\sigma}\right\}\right\},$$

$$u_2 = k_2 N_2 \max\left\{0, \min\left\{1, \frac{N_2 - M_2 + \sigma}{2\sigma}\right\}\right\}$$

(19.14)

where $k_1 > \sup \varepsilon_1$ and $k_2 > \sup \gamma_{21}$. In this case, $\dot{N}_1 < -\varepsilon_2 N_2 < 0$ if $N_2 \geqslant M_2 + \sigma$, for which reason the condition $N_2 \leqslant M_2 + \sigma$ appears in due course and remains valid. From above it follows that $\dot{N}_1 \geqslant N_1(\varepsilon_1 - \gamma_{12}(M_2 + \sigma))$ at $u_1 = 0$ and N_1 increases in the domain $N_1 \leqslant M_1 - \sigma$ if the initial value of N_1 is not zero. But if $N_1 \geqslant M_1 + \delta$, then $u_1 = k_1 N_1$ and $\dot{N}_1 \leqslant -N_1(k_1 - \varepsilon_1)$, and so the image point of the system turns out to be within the band $|N_1 - M_1| \leqslant \sigma$ after a time. Here, when $N_2 \leqslant M_2 - \sigma$, the coordinate N increases if its initial value is not zero. Therefore,

$$|N_2 - M_2| \leqslant \sigma$$

after a lapse of time.

Thus, if the initial values are $N_1 \neq 0$ and $N_2 \neq 0$ in the system (19.13), (19.4), then, after a time, the image point (N_1, N_2) proves to be in the square $2\sigma \times 2\sigma$ with its center at the point (M_1, M_2) for any arbitrary preassigned value of $\sigma > 0$. The plot illustrative of a qualitative

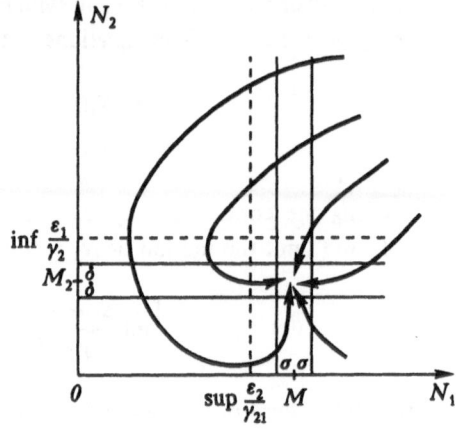

Fig. 19.9.

character of the system trajectories is shown in Fig. 19.9.

20. Power Engineering: Control of a Nuclear Reactor

An atomic power plant produces electric energy from heat generated in a nuclear reactor that must be under control so as to ensure its safe operation and generate more power. We will consider a control problem with the aid of the so-called lumped model intended to represent the reactor. This model only conceptually reflects the processes occurring in the nuclear reactor, and so the results outlined below are directly unsuitable for atomic power plants.

Sections 20.1 through 20.4 contain the most elementary information on atomic energy and nuclear reactors. The reader experienced in the issues involved can at once pass on to Sec. 20.5 which includes equations of the model.

20.1. Why Does the Nuclear Reactor Heat up?

A nuclea reactor, like an atomic bomb, releases the binding energy of atomic muclei of a special material — a nuclear fuel. The fuel includes isotopes of some heavy elements, in particular, uranium-235. Each uranium isotope contains in its nucleus 235 nucleons (protons and neutrons). Natural uranium approximately consists of 0.7% of these isotopes, and the rest of isotopes are those of uranium-238.

The binding energy is released in the fission reaction of nuclei bombarded by neutrons. As a neutron falls into a nucleus, the latter decays into two fission fragments that emit a few neutrons and a few photons. The fragments undergo a number of acts of the β-decay with the emission of electrons, photons, and antineutrinos. As a result, there appear almost everywhere stable daughter (product) nuclei, although in some cases the fission fragments also emit delayed neutrons after the process comes to a halt. The total energy of the fission fragments and of the resultant particles appreciably exceeds the energy of the electron that triggers the reaction (by ten orders of magnitude, in some cases).

A major share of the fission reaction energy (about 80%) is the kinetic energy of scattering fragments which slow down in the environment and heat it. In the final analysis, the energy of almost all energy-emitting particles also converts to heat, with the exception of the energy of antineutrinos: these particles do not actually interact with the substance and freely leave the reactor, carrying away about 5% of fission energy.

To heat up one gram of water by one degree Celsius requires the same amount of energy as that released during the fission reaction of 1.3×10^{11} nuclei. Therefore, many billions of billions of nuclei must undergo the fission at every second in the nuclear reactor of an atomic power plant. It takes a great quantity of neutrons to bombard these nuclei. Where are the neutrons to be found?

20.2. Chain Reaction

In the fission reaction, a neutron falling on a nucleus releases a few neutrons which are, in principle, able to cause new nuclear fissions attended by the appearance of new free neutrons, and so on (Fig. 20.1). This process is said to be the chain reaction of nuclear fission, in which the generations of neutrons give way to each other in the same fashion as the generations of living beings: a "parent" neutron ceases to exist as it collides with a nucleus, but leaves a few "descendants". The course of the chain reaction depends on the reproductive capacity of a neutron population. If the posterity of one neutron includes, on the average,

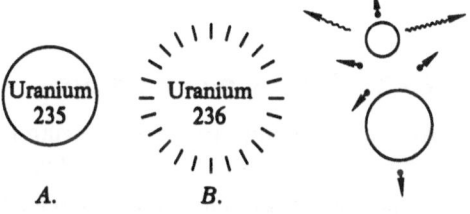

Fig. 20.1. The fission reaction of a nucleus (A) of uranium-235 can arise from the collision of a neutron with the nucleus. In approximately one case out of 70, the electron may bounce from the nucleus. In the rest of the cases, the nucleus may absorb the neutron. This leads to the formation of a nucleus (B) of uranium-236, which is in the excited energy state. The latter nucleus can pass to the ground energy state due to the emission of photons. But in six out of seven cases, the nucleus (B) undergoes the fission into two fission fragments accompanied by the emission of photons and neutrons.

one fertile specimen, the size of the neutron population remains approximately constant. If the number of reproductive descendents is less than one, on the average, then the reaction comes to a halt, but if the number is higher, the reaction rate increases in the avalanche-like fashion. Note that the neutron population "explosion" can lead to a natural explosion.

For the nuclear reactor to be able to produce a fairly high amount of heat, it is necessary to work out the conditions for the initiation and development of the chain reaction. In this case, in the reactor must appear free neutrons capable of breeding a vigorous population the size of which must first grow on and then become stable at the prescribed level. What is the way of achieving these conditions?

Let us note that the original substances for the generation of neutrons are freely available in nature. Free neutrons are always present in the mass of a nuclear fuel. Some neutrons come to the Earth together with cosmic rays and other appear in uranium during the spontaneous fission of its nuclei; in one kilogram of natural uranium, about 15 neutrons appear every second.

In order that these neutrons should produce a self-sustained chain reaction, specific conditions are necessary because neutrons are not too readily reproducible. The only example is known of the emergence of these conditions in nature: in the uranium deposit of Gabon, the chain reaction lasted 600 thousand years, but still terminated two billion years ago. What does the reproductive capacity of a neutron depend on? To understand the cause, we will take a closer look at how it exists.

20.3. Existence of a Neutron

As it moves within a substance, a neutron encounters on its way atomic nuclei from time to time and gives rise to different nuclear reactions. The collision of the neutron with a nucleus is a random event, and the probability of one or another of the nuclear reactions depends both on the kinetic energy of the neutron and, naturally, on the kind of nucleus the neutron collides with. We will not enumerate all possible reactions involved with neutrons and limit the discussion to the reactions caused by fission neutrons (released in fission reactions) in the fuel core of the reactor. Let us consider a uranium reactor using the mixture of uranium-238 and uranium-235 as a fuel.

A statistical average fission neutron exhibits an energy that comprises about one percent of the entire energy liberated during the fission reaction. This type of energy neutron is said to be fast. As it runs into a nucleus of any uranium isotope, the fast neutron can initiate the fission reaction. But it is more probable (by about an order of magnitude) that the neutron will be subject to the inelastic scattering and rebound from the nucleus. In this case, a major share of the neutron energy passes to the kinetic energy of the nucleus or to the energy of photons emitted by the nucleus, so that the neutron will completely lose the ability to cause the fission of nuclei of uranium-238. If this slow neutron meets with a nucleus of the above isotope, then the most probable reaction will be the reaction of the radiative capture of the neutron followed by the emission of photons. The neutron will decay without leaving descendants. It is only uranium-235 that offers this neutron an opportunity for the reproduction. Therefore, to perform the fast neutron chain reaction, it is necessary to secure a high probability of the collision of neutrons with nuclei of uranium-235. But this reaction calls for highly enriched uranium having a content of uranium-235 of 20 to 30%. This fuel is rather expensive, and the design of fast reactors and their service are fairly complex. For this reason, these nuclear reactors, now available all over the world, provide only less than one-tenth the entire amount of electric power produced at atomic power plants.

Reproduction of neutrons in a chain reaction is shown in Fig. 20.2.

Most of the nuclear-power reactors use weakly enriched uranium (up to 5% of uranium-235) and operate on slow thermal-velocity neutrons, which go under this name because they remain in the state of the thermal equilibrium with atoms of the medium. The most probable outcome of the collision of a thermal neutron with a nucleus of uranium-238 is the elastic scattering similar to the recoil of elastic balls in classical physics. But the collision with the nucleus of uranium-235 almost always results in the fission reaction, which is what is necessary.

For it to be thermal, a fission neutron must decrease its energy by eight orders of magnitude, on the average. How can one ensure such an appreciable neutron thermalization? We noted above that the moderation of a neutron in uranium is highly improbable. Therefore, special moderators of neutrons are fitted into nuclear reactors. Water or graphite mostly serves as a moderator the atomic nuclei of which more readily scatter neutrons than capture them. To reduce the absorption of neutrons in uranium, the reactor fuel is divided into small portions dis-

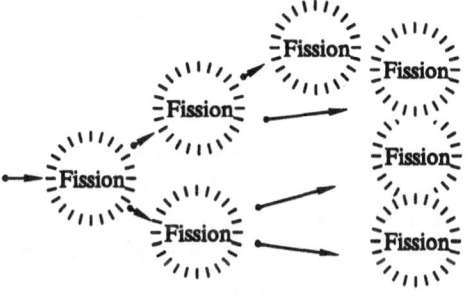

Fig. 20.2. Reproduction of neutrons in a chain reaction occurs in the following fashion: neutrons that appear during the fission reaction trigger new fission reactions, and so on. The diagram illustrates an avalanche-like development of the chain reaction, which is specific to an atomic bomb rather than to a nuclear reactor. In a nuclear reactor, less than one-half of the fission neutrons give rise to new fission reactions and the remaining neutrons are absorbed by nuclei in the radiative capture reactions or escape from the fuel core. The diagram also illustrates the reactions in which nuclei scatter neutrons: a neutron collides with and recoils from the nuclei before it commonly induces the fission reaction.

posed in what are known as fuel elements (heat-releasing elements) and enclosed by moderators. As it leaves a uranium piece, where it came into existence, a neutron penetrates into a moderator, gradually loses its velocity, and passes in due course to an irregular motion of a thermal Brown particle. While wandering within the moderator, the neutron may again appear among uranium nuclei where it has a good chance of causing the fission reaction and leaving the posterity, namely, an average of 2.5 fast neutrons.

If the fission occurs and new neutrons appear, these neutrons repeat the way of life of their parent: first, they slow down and reach the state of the thermal equilibrium with the medium, randomly move in this medium, and finally die out in the fission reaction of the nucleus of uranium-235. In this way of its life, a neutron may run two risks: it can move beyond the fuel core of the reactor or undergo the radiative capture. Potential acceptors are certainly not only nuclei of the fuel and moderator, but also any nuclei found to be in the fuel core of the reactor. Apart from different admixtures, the fuel core is sure to include a heat carrier, a structural material, and fission reaction products. Besides, the reactor contains a substance specially placed into it for a vigorous absorption of neutrons. This is done with the aim to change the amount of the absorber in the reactor, i. e., to increase or decrease the probability of the neutron absorption, and hence to control the course of the chain reaction — to preclude the elimination and the explosion of the neutron population.

An individual success of the reproduction or a failure of the reproduction of each neutron depends on a set of random factors. But if we consider the entire population of neutrons in the reactor, the sum of these factors will now appear to be a determinate factor of a certain character of the chain reaction, which depends on the set of the physical and chemical conditions of the nuclear reactor and on the reactor design.

20.4. Populations of Neutrons

Assume that the sizes of two successive populations in the fuel core of a nuclear reactor are N and $N + N$, respectively. The number N can naturally have any sign or can be equal to zero. The operation of the reactor can be defined by the factor $\rho = \Delta N/(N + \Delta N)$, called the reactivity. Its value specifies the course of the chain reaction and reflects the entire set of the existence conditions of neutrons. If < 0, the chain reaction decays; if $\rho > 0$, the reaction grows in intensity; and if $\rho = 0$, the steady operation sets in.

Using the above factor, it is easy to describe the progression of the chain reaction in time:

$$\frac{N(t) - N(t - l)}{l} = \frac{\rho}{l} N(t)$$

where $N(t)$ is the number of neutrons at the instant t and l is the mean lifetime of the population. The above equality can be replaced at a low error by the equation

$$\dot{N} = \frac{\rho}{l}N \qquad (20.1)$$

The reactivity accounts for the probability ρ of both the escape of a neutron from the fuel core of the reactor and its radiative capture within this core. The probability depends on the reactor design, substances used, enrichment of the fuel and fission products present in the fuel, current temperature, heat-release power, heat-carrier pressure, etc. To change the reactivity in an effective way and hence to control the course of the chain reaction, the design of any reactor stipulates devices which make it possible, if required, to increase or decrease in the fuel core the amount of the absorber — the substance containing atoms the nuclei of which accomplish the radiative capture of neutrons at a high probability. These devices commonly take the form of retractable rods; boron or cadmium serves as an absorber.

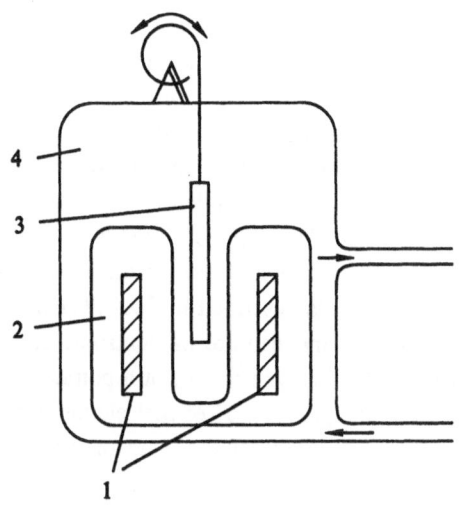

Fig. 20.3. The schematic diagram of a thermal reactor in section: 1, fuel elements; 2, moderator; 3, absorbing rod; and 4, heat carrier. The fuel elements contain a nuclear fuel, which is commonly uranium or its oxide. Water is most often used as a heat carrier. Graphite or water that perform the function of the heat carrier serves as a moderator. The absorbing (control) rod contains boron or cadmium and serves to change effectively the reactivity.

Nuclear reactors are designed so as to obtain both a positive reactivity on pulling the absorbing rods completely out of the fuel core and a negative reactivity on pushing the rods completely in the fuel core (Fig. 20.3). In addition, designers envisage the interconnection between the reactivity and the power and temperature through negative feedback loops: when the rods are stationary, the reactivity decreases if the temperature or the heat-release power increases in the working range of variations of these parameters. Finally, the reactivity diminishes with time because the nuclear reactor fuel elements exhibit a decrease in the amount of uranium-235 and accumulate the fission products prone to the radiative capture of neutrons.

20.5. Mathematical Model of the Reactor

The dimensions of power reactors are always appreciable. The thing is that the share of neutrons that escape outside from a larger fuel core is less than that of neutrons escaping from a smaller core because a large volume has a relatively small boundary surface. The fuel core of a typical reactor reaches hundreds of cubic meters and contains tens of thousands of fuel elements filled with hundreds of tons of uranium. In such a large structure, the chain reaction cannot proceed uniformly over the entire volume; the same is also true for all attendant processes. The nonuniformity is responsible for the processes of the scatter of energy and particles over the reactor volume.

An adequate pattern of the above events in a mathematical model can be set up only after a detailed study of the specific structure of the reactor under investigation. The result of this study will most likely be a very complex model. To simplify the discussion, we diregard the above nonuniformity and assume that all physical events described in Secs. 20.1 through 20.4, such as the nuclear fission, moderation and capture of neutrons, and heating of fuel elements, occur in a negligible volume or simply at a point.

The temperature T of fuel elements and heat-release power W are taken to be the coordinates of the state of the reactor under study. The fuel elements are heated by way of the transfer of energy to the heat carrier. The phenomena involved are defined by the equation

$$M\dot{T} = W - k(T - \overline{T}) \qquad (20.2)$$

where M is the heat capacity of fuel elements, T is the heat carrier temperature; and k is the heat transfer factor. We suppose that k, M, and T are constants whose values are known.

Each fission reaction releases and converts to heat a certain constant amount of energy. Consequently, the heating power W of the reactor is proportional to the current rate of reactions. The number N of neutrons is also proportional to this rate. In view of (20.1), we obtain

$$\dot{W} = \frac{\rho}{l}W \qquad (20.3)$$

where $l = $ const is the mean lifetime or the reactivity of a neutron. Equations (20.2) and (20.3) form the model of interest.

As noted in Sec. 20.4, the reactivity depends on W, T, t and the position of control cores which we specify by the quantity U. Thus, $\rho = \rho(T, W, U, t)$. Let us expand the function ρ into a power series near an arbitrary point T_*, W_*, U_* within the working range of changes in these quantities:

$$\rho = \rho(T_*, W_*, U_*, t) + \frac{\partial \rho}{\partial T}(T - T_*) + \frac{\partial \rho}{\partial W}(W - W_*) + \frac{\partial \rho}{\partial U}(U - U_*) + \dots$$

Assume that we can disregard all cofactors of the power higher than the first, except for the cofactor

$$\frac{1}{2} \frac{\partial^2 \rho}{\partial T^2} (T - T_*)^2.$$

Specialists admit that this assumption does not contradict reality; to be more specific, it does increase the difference between the reactor and its lumped model. Thus, it can be written that

$$\frac{1}{l} \big(\rho(T, W, U, t) - \rho(T_*, W_*, U_*, t) \big) =$$

$$= a(T - T_*) + b(W - W_*) + c(U - U_*) + d(T - T_*)^2 \qquad (20.4)$$

Assume that a, b, c, d are variable, bounded, nonmeasurable parameters; $c = c(t) >$ const > 0; and equality (20.4) holds for all values of the arguments.

20.6. Statement of the Problem and Properties of the Model

Consider a control problem in which we need to change the quantity U so that the image point (T, W) of the system (20.2), (20.3) tends to a certain prescribed state (T^*, W^*). Assume that we bear in mind just this prescribed state in the expansion (20.4). The pair (T^*, W^*) cannot be chosen arbitrarily; we assume that this pair reduces T^* and W^* to zero at a certain value of U. Similar constraints on the choice of the prescribed quantities were met with in Sec. 19.

We introduce the designations: $x_1 = T - T_*$, $x_2 = W - W_*$, $u = U - U_*$, and $r(t) = \rho(T_*, W_*, U_*, t)$. Considering that $W_* - k(T_* - \overline{T}) = \dot{T}|_{T=T_*} = 0$ and allowing for (20.4), we can represent the model as

$$\dot{x}_1 = \frac{1}{M} (x_2 - kx_1)$$

$$\dot{x}_2 = \big(r(t) + a(t)x_1 + b(t)x_2 + d(t)x_1^2 + c(t)u \big) (x_2 + W_*) \qquad (20.5)$$

From the conditions $T > 0$ and $W \geqslant 0$ it follows that the space of states (x_1, x_2) of the system (20.5) should be thought to be one-fourth of the plane $x_1 \geqslant -T_*$, $x_2 \geqslant -W_*$. In these coordinates, the problem involves the construction of a control that causes the system to move to the coordinate origin (Fig. 20.4).

According to (20.5), the straight line $-x_2 = W_*$ is an invariant set over which the system moves to the state $x_1 = \overline{T} - T_*$, $x_2 = -W_*$. This corresponds to the obvious property of the extinct reactor: its temperature graduallu approaches the temperature of the heat carrier. But the invariance of the above straight line points out that it is impossible to regenerate the extinct reactor, which is contrary to facts. The error in the system (20.5) comes from equation (20.1) which, despite the facts, does not admit the existance of "native" neutrons emerging without the performance of the chain reaction. For the model to be consistent with reality, we assume that the initial state of the reactor obeys the condition $x_2 > -W_*$.

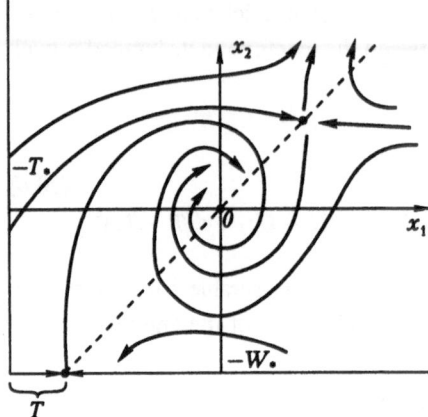

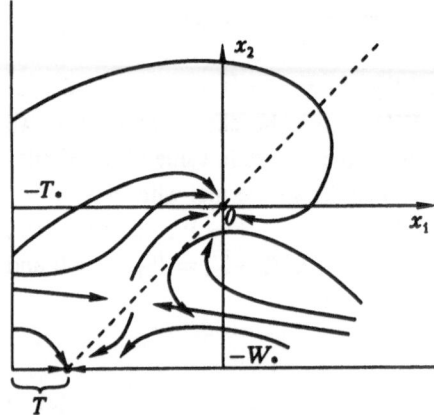

Fig. 20.4. Behavior of the reactor model with constant parameters and constant control. The phase portraits relate to "well designed" reactors for which the temperature and power coefficients of the reactivity are negative. These coefficients are the parameters a and b in (20.5); at the left, $d > 0$ and at the right, $d < 0$. In both of the cases, the zero solution is locally stable, but there are states in which it is impossible to approach the coordinate origin.

In view of the first equation of (20.5), singular points of the system lie on the straight line $x_2 = kx_1$. Their number and location depend on values of the parameters a, b, c, d, r and the control u. When

$$u = -\frac{r(t)}{c(t)}$$

the coordinate origin becomes a singular point. In the general case, however, this equalibrium position is not stable in the "large". Indeed, if all the parameters are constant and $d \neq 0$, then the system includes a singular point with coordinates

$$x_1 = -\frac{a+kb}{d}, \qquad x_2 = -\frac{a+kb}{d}\,k;$$

it is evidently impossible to move from this point to the coordinate origin. Two examples of the phase portraits of the system (20.5) are given in Fig. 20.4.

The interrelation stated in Sec. 20.4 between the reactivity, temperature, and power is defined in the model (20.5) by the parameters $a < 0$ and $b < 0$ at values of x_1 and x_2 that are close to zero. But we will not consider this circumstance anywhere in the text that follows. For it to be taken into account, we would have to describe the dependence of a and b on the system coordinates because at high values of x_1 and x_2, the sign of the coefficient can be different. This aspect is not worth the trouble because, as will be seen from the subsequent discussion, the problem lends itself to the solution without the information on this interrelation. Of course, when the case in point is an actual nuclear reactor where there is a need for saving a control action, it is evidently worthwhile to estimate the usefulness of this information rather than to disregard it at all. But in our case, we examine an idealized system, although the parameters a and b of the model can be positive at any x_1 and x_2.

20.7. Why Must the Control be Nonlinear?

Even at first sight, the linear control law cannot impart the system the desired properties. Indeed, at constant parameters of the model, the control $u = k_1 x_1 + k_2 x_2$ only changes values of the coefficients attached to the first-order terms in the right sides of the equations of the system, but does not affect the quadratic reactivity term which just generates a singular point beyond the coordinate origin. However, it would be desirable to be sure that this type of control is groundless in a more general case, too.

Let us prove the inefficiency of any control of the form

$$u = k_0(x,t) + k_1(x,t)x_1 + k_2(x,t)x_2 \qquad (20.6)$$

with bounded functions k_0, k_1 and k_2 for the system (20.5) under the condition $\inf_t d(t) > 0$. From the first equation of (20.5) it follows that $\dot{x}_1 \geqslant 0$ when $x_2 \geqslant kx_1$. Therefore, the image point of the system can leave the angular domain of the form $x_2 \geqslant kx_1$, $x_1 \geqslant X = \text{const}$ only through a sloping boundary, i.e., through a segment of the straight line $x_2 = kx_1$. In view of (20.5) and (20.6), at all points of this straight line we have

$$\dot{x}_2 - k\dot{x}_1 = \dot{x}_2$$
$$= \left[r + ck_0 + (a + ck_1)x_1 + (b + ck_2)x_2 + dx_1^2\right](x_2 + W_*)$$
$$= \left[r + ck_0 + (a + ck_1 + bk + ckk_2)x_1 + dx_1^2\right](x_2 + W_*)$$

If $x_1 \geqslant X$ at a fairly high X, the above expession in the square brackets is positive, and so the quantity $x_2 - kx_1$ grows in magnitude. Consequently, the system cannot escape from the domain $x_2 \geqslant kx_1$, nor can it approach the coordinate origin.

The inefficiency of the control (20.6) proves that there is a need to introduce the terms of the order higher than the first into the control. It is certainly not evident from the above that we need to use only just the binary induction control. However, it would be unreasonable to neglect the control synthesis technique suitable for a wide class of problems, especially in the case when the objective of this chapter is to illustrate this technique.

20.8. Design of the Control Law

According to the technique developed in this course book, the control problem of the object (20.5) can be replaced by two successive problems: the first problem should enable us to determine the induction error $\sigma(x_1)$ and the operator $v(x_1)$ of internal feedback the generation of which ensures the desired character of motions and the second problem should make it possible to derive the induction action $u(x)$. What form must the functions σ and v have? Note that the nonmeasurable free term $r(t)W_*$ is present in the right side of (20.5). Therefore, it is necessary that the condition $\sigma(x_1) > 0$ should be valid for all x_l. For simplicity, we set $(x_1) = \text{const}$. The induction of requisite feedback at an error of σ, is defined by the equality $x_2 = v + \mu\sigma$ under the constraint $|\mu(t)| \leqslant 1$. In view of the first equation of the system, we then have

$$\dot{x}_1 = \frac{1}{M}\left(v + \mu\sigma + kx_1\right) \tag{20.7}$$

Equation (20.7) represents a model of the one-dimensional system with the output x_1 and input v, for which we need to solve the control problem. If we define the input v by linear feedback, $v = hx_1$, then at $h < k$ the point $x(t)$ will tend to the neighborhood of zero of the radius $\sigma/(k - h)$. This conclusion obviously follows from the equation of the closed system

$$\dot{x}_1 = \frac{h - k}{M}x_1 + \frac{\mu\sigma}{M}$$

On decreasing h, it is possible to reduce the radius of the above neighborhood and thus speed up the approach of the point to the neighborhood. The second circumstance is more important than the first because the neighborhood radius is easy to control through the choice of σ.

The linear operator v suffers from the following drawback: its plot is put inside the phase space of he system only if $0 \leqslant h \leqslant W_*/T_*$, but at other values of h there exist values of $x_1 > -T_*$, such that $hx_1 < -W_*$, and the equality $x_2 = v$ is impossible to achieve. Therefore, we should either select h from the section $[0, W_*/T_*]$ or use a nonlinear function. The first version does not look too attractive because it eliminates the "fast-running" negative values of h. Let us define the nonlinear function.

$$v(x_1) = \max\{hx_1, -W_0\}$$

where $0 < W_0 < W_* - \sigma$. It is clear that $v = hx_1$ at $h \in [0, W_0/T_*]$. From the first equation of (20.5) it follows that the motion near the broken line $x_2 = v(x_1)$ at $h < 0$ ensures a faster decrease in $|x_1|$ than the motion near the straight line

$x_2 = v(x_1)$ at $h \geqslant 0$. Let us now derive the control u that ensures after a certain time the inequality $|s| \leqslant \sigma$, where $s = x_2 - v(x_1)$. We have

$$\frac{d}{dt}|s| \leqslant \dot{x}_2 \operatorname{sgn} s + |h| \cdot (1/M)|x_2 - kx_1| \leqslant$$

$$\leqslant |r + ax_1 + bx_2 + dx_1^2|(x_2 + W_*) + (|h|/M)|x_2 - kx_1| + c(x_2 + W_*)u \operatorname{sgn} s$$

Let $A \geqslant \max\{r(t), a(t), b(t), d(t)\}$. Recall that $\inf c(t) > 0$. Then, at $\operatorname{sgn} u = -\operatorname{sgn} s$ and

$$|u| > \frac{1}{\inf c}\left[A(1 + |x_1| + |x_2| + x_1^2) + \frac{|h|}{M}\frac{|x_2| + k|x_1|}{x_2 + W_*}\right]$$

the quantity $|s|$ decreases in magnitude. We need this type of motion only when $|s| \geqslant \sigma$. Therefore, in the band $|s| \leqslant \sigma$ we will decrease the control so as to avoid its discontinuity on the line $s = 0$. For this, it is possible to introduce, for example, the cofactor $\min\{1, s/\sigma\}$ into the control law. Finally, the control law assumes the form

$$u = -H \min\left\{1, \frac{|s|}{\sigma}\right\}\left[A(1 + |x_1| + |x_2| + x_1^2) + \frac{|h|}{M}\frac{|x_2| + k|x_1|}{x_2 + W_*}\right]\operatorname{sgn} s$$

$$(20.8)$$

where $H > 1/\inf c(t)$. The behavior of the closed system in regard to its quality is shown in Fig. 20.5.

20.9. Properties of the Closed System

The control (20.8) causes the image point of the system to penetrate into the band $G = \{x : |x_2 - v(x_1)| \leqslant \sigma\}$ and remain within it. The residence of the point in this band entails the motion of the system to the coordinate origin, at least until the point $x(t)$ appears in the domain

$$|x_1| \leqslant \frac{\sigma}{k - h}, \quad |x_2| \leqslant \frac{k + 2|h|}{k - h}$$

Selecting a sufficiently small value of σ, we can achieve any desired accuracy of the system stabilization.

The coefficient H determines the speed at which the system reaches the band G from its initial state: the higher the value of H, the faster the system moves. As the point $x(t)$ falls into G, the character of its motion and changes in $u(t)$ depend little on H; the parameters h and W_0 begin to play an appreciable role. As noted above, at relatively high values of $|h|$ and W_0, chosen at $h < 0$, the system moves faster to the coordinate origin, which is likely to be desirable.

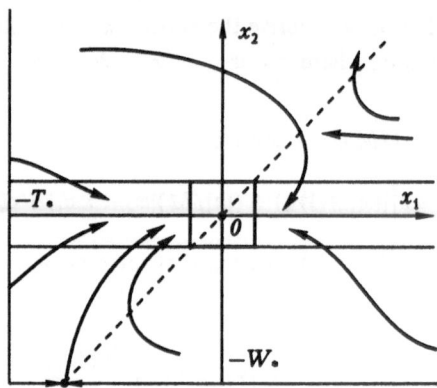

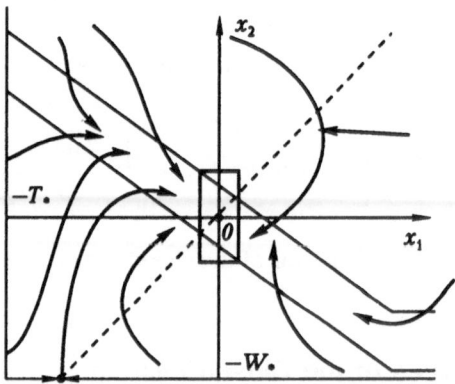

Fig. 20.5. In the closed system under any initial conditions, excepting $x = w$, the trajectory $x(t)$ reaches the band $|s| \leqslant \sigma$ in a finite time, approaches along this band the coordinate origin, and finally finds its way into rectangles the sizes of which can be made as small as desired by way of the requisite choice of σ. The upper and lower figures correspond to the parameter $h = 0$ and $h < 0$, respectively.

However, the motion from the domain $x_1 < 0$ (i. e., at a temperature of a fuel element that is below the specified one) must proceed at values of $x_2 > 0$ that correspond to an increased heat-release power. An increase in the power may appear dubious in the eyes of the workers who attend real nuclear rectors; in this connection, they will possibly prefer to keep the parameter h at zero. The control system with h held at zero moves from the initial state onto the band $|x|_2 \leqslant \sigma$ and then slowly attains the prescribed temperature near the desired power level. It may be that when $x_1 > 0$, the behavior of the system at $h < 0$ looks more attractive: the overheated reactor ($x_1 > 0$ points to an increased temperature) has its power decreased to a level of $W_\bullet - W_0$ (i. e., to a value of $x_2 = -W_0$), with the result that it rapidly cools down and raises its power to the specified one, but now at a safe temperature when $x_1 \approx 0$.

The above advantages for various versions of the choice of parameters can be united by prescribing a new function

$$v(x_1) = \min\{0, \max\{hx_1, -W_0\}\}$$

In this case, the control law (20.8) is suitable; however, it should be kept in mind that the quantity s is $s = x_2 - v(x_1)$ at the new v law.

21. Economy: Stabilization of the Trajectories of Growth

A government commonly has possibilities of exerting actions on the economy of its country. Examples of the actions are familiar to all of us: long-term plans, protectionism, rationing of products, etc. However, in their economic activities, authorities are not bound to interfere in the mechanism of functioning of a free market. The text of this section shows that organs of government can be of benefit even to a sagging market of the doomed economic structure.

21.1. On Modeling of Economic Systems

At first sight, economics today is found to be a fully mathematized branch of science. Books on economics are overfilled with digits, tables, formulas, and other criteria of genuine scientific nature. However, a closer scrutiny verifies that a huge number of these mathematical features define only the current values and the interrelation between some economic indices at individual instants of time. The course of a certain process in economics received very little study. The system approach is not yet often used to carry out economic investigations, and it is by no means easy to extract from a great body of scientific literature an acceptable model of the dynamic system relating to the economic subject. The real cause of this disappointing situation is that the phenomena of interest are rather complex and the scope of possibilities for their study is fairly small. Indeed, the life of people from the economic viewpoint depends on the joint behavior of large number of persons whose actions are not easy to analyze. The development of behavioral models generally relies on unsteady reasons of common sense. The check of these models for adequacy appears to be inconclusive because the researcher has to do with the real system (national economy) of the only specimen the experiments on which are unfeasible. One has to be content with a model that displays a plausible behavior under input actions resembling real actions.

Recall that the theory of automatic control clarifies the ways of changing the input actions so as to change the system behavior in the requisite way. But if we change an input, the behavior of the economic model may vary in a way other than the way of changes in the behavior of the economy itself. Therefore, the control algorithms corresponding to the present economic models cannot be of practical significance. However, we will have to hope that the development of these algorithms gives impetus to economists to go to the heart of modeling, so that in a certain time these algorithms will raise for cyberneticists such control problems which actually relate to practice.

The control problem at hand is devoid of direct practical importance. We will consider and ensure the stabilization of the exponential growth of market economy only for a rather imperfect model. Therefore, our structures do not abolish in any way the sword of Damocles involved with the historic process and do not deprive the grave-digger of bourgeoisie the hope to reforge in due time this sword into a spade and, hence, to fulfil the historic mission.

21.2. Essence of Processes Under Study

The economy of any type (market or planned economy) deals with the production and consumption of a diversity of products. A portion of the products are intended for ultimate consumers: these are goods and commodities used to satisfy the wordly needs of people, which range from the daily bread to articles of refined luxury. There is reason to assign to the ultimate consumers their government, too, because it acqures means to meet the demands of the population for its protection (army), freedom (parliament), order (prisons), and so on. Another portion of products represent means of production, which serve to increase the fixed capital.

We will resort to a model that describes variations in three quantities (induces): the production, consumption (expenditure), and capital.

The difference between the planned and market systems lies in the character of changes in the above quantities. In the planned economy, they are subject to the tradeoff between the requirements of directive bodies and to production capabilities; the study of this tradeoff is beyond the scope of our problem. In the market economy, the indices of the national economy conform to the aggregate of analogous characteristics for a large number of firms, in which case the economic indices of each firm depend on the conditions of the market. For example, if the solvent demand for the output of a certain firm is high, the firm raises the output of products to the best of its productive capacities (fixed capital), and the owners of capitals strive to invest means for expanding these capacities.

The market involves trade, money, prices of products, labor costs (wages), and loans (rates of interest). For simplicity, we do not examine these market features, assuming that the market conditions depend on the three factors (quantities) mentioned above: the production, consumption, and capital. But we will refer to some features of commerce all the same because we need to describe quantitatively the economic activity. Here, it is necessary to use universal units of measurement, which are equally applicable both to goods and services for ultimate customers and to means of production. It is quite natural to measure the quantity of production by its value expressed in certain monitory units at fixed prices, for example, in the US dollars at prices as of 1989. Thus, the behavior of the national economy can be defined in terms of the three varying numerical parameters (Fig. 21.1).

21.3. Equations of the Model

Let K identify the current cost of the fixed capital, C denote the current consumption per unit time, and Y denote the current volume of production (output) per unit time, which is left after the subtraction from the gross national product the means required to compensate for the reduction of the fixed capital. Recall that in the process of production, there occurs a depreciation of the fixed capital, i.e., the wear of equipment, so that it is necessary to make up constantly for the loss of the capital.

We describe changes in these quantities by the equations

$$\dot{K} = \alpha Y - \beta K$$
$$\dot{Y} = \lambda(C + \dot{K} - Y) \tag{21.1}$$

where α, β, and λ are positive parameters. We will interpret in detail these equations.

The first equation of (21.1) describes the policy of investments. It seems reasonable to accept that for any output Y there exists a definite value of the fixed capital K, which is optimal in a certain sense for this output. The model presupposes that the optimal capital is proportional to the output and the capital investments are proportional to the deviation of the actual fixed capital stock from the optimal one. Satisfying a high demand for products, firms can ensure a high level of the output Y at a small volume of the capital K by recruiting an additional labor. But it is more profitable for the firms to acquire additional facilities and increase K. However, to raise the fixed capital in excess of the necessary volume can entail the idle time of some equipment, which reduces the profit.

The second equation of (21.1) reflects the policy of firms in the area of production. On replenishing the fixed capital, we express the output of products by the quantity Y and the amount of marketing of the products by the sum $C + K$; comparing these quantities, a firm takes a decision to raise or cut down

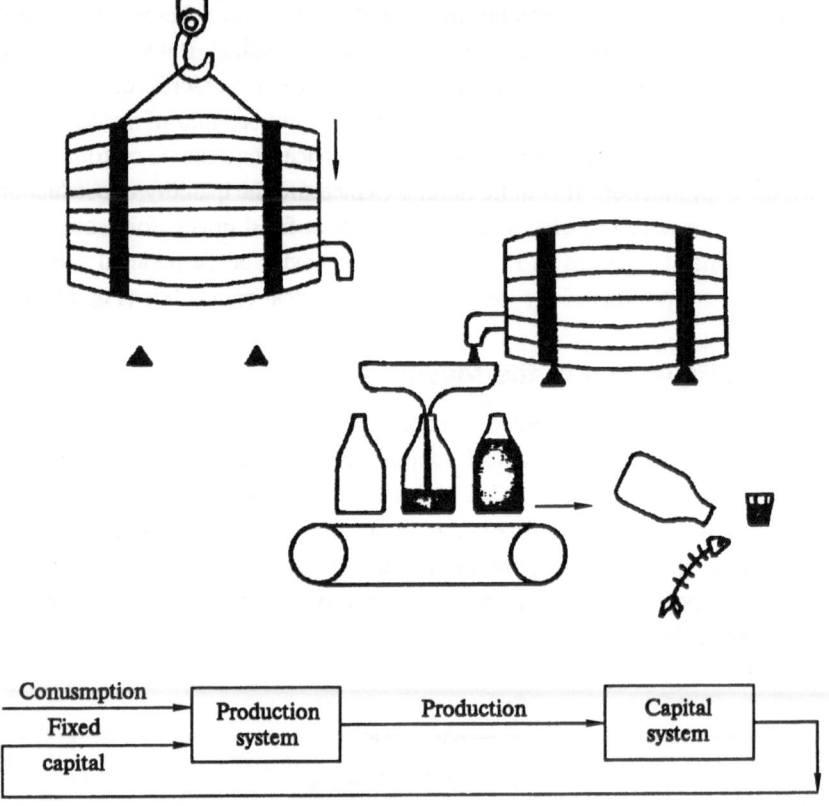

Fig. 21.1. The representation of economy as an aggregate of dynamic systems gives grounds to recall the historical problem as to what appeared earlier: a hen or an egg? The discussion of the fact that economy offers "everything for the welfare of mankind" can involve the statement of the following interrelation between economic processes (the upper figure): changes in the fixed capital (production installation 1) causes changes in the volume of output (2), which tells on consumption (3). But the analysis of the operation of economy calls for studying the behavior of firms that look after their own welfare. Firms interrelate the above processes in another way: they respond to variations in consumption (3) by changing the volume of output (2), which stimulates the firms to change the volume of fixed capital (1). Therefore, the national economy as an aggregate of firms can be depicted in the form of the pattern presented in the lower figure.

the output. Using the first equation of the system (21.1), we can omit K from the right side and obtain the notation of a more customary form:

$$\dot{Y} = \lambda(C + (\alpha - 1))Y - \beta K.$$

The quantity C is the input variable of the dynamic system (21.1). We use C as a control. The overall final consumption results from the individual consumption by the population and from the government expenditures for goods and services. The government can change its expenditures within certain limits at the discretion of the parliament that approves the state budget. Our further discussion can help the government that deals with the economy of the form (21.1) to pursue such a budgetary policy that would contribute to the economic prosperity.

21.4. Labor Force and Constraints on Variables

If a government incurs some very large expenditures required, for example, to implement costly social, cosmic, or military programs, the quantity C becomes relatively high in magnitude, which, in view of (21.1), causes an increase in Y and K. Judging from the equations, the growth of the government expenditures can spur the economic progress as fast as desired. This improbable conclusion attests that equations (21.1) alone are not sufficient for the description of the economy.

Equations (21.1) disregard the fact that the production calls for the labor force apart from the fixed capital. An increase in the size of the labor is retarded, at least by the demographic circumstances, for which reason the output of products and the volume of the fixed capital cannot grow too fast. It is common to suggest that the size L of the labor required to turn out products Y with the availability of the fixed capital K is given as

$$L = Be^{-\rho t}Y^b K^{1-b} \tag{21.2}$$

where $b > 1$, $B > 0$, and $\rho > 0$. The mappings of this type are known as Cobb-Douglas production functions. The cofactor $e^{-\rho t}$ accounts for the technical progress: the human labor expenditure for a certain output of products at the prescribed fixed capital exponentially decays with time.

Assume that the size of the workable population exponentially increases:

$$L_* = L_0 e^{lt} \tag{21.3}$$

The natural constraint on the size of labor has the form

$$L_* = L_* \tag{21.4}$$

We have reason to hold that the quantity Y is defined by the second equation of the system (21.1) when (21.4) is a strict inequality; otherwise Y is given by the equality

$$Y = \left(\frac{L_0}{B} e^{(\rho+l)y} K^{b-1}\right)^{\frac{1}{b}},$$

which is the consequence of (21.2) and (21.3). Note that the second case corresponds to the full employment of the population.

In addition to (21.4) and the obvious conditions $Y \geqslant 0$, $K \geqslant 0$, and $C \geqslant 0$, the model requires constraints accounting for the elementary fact that one cannot consume products if they are unavailable. Strictly speaking, we should examine the dynamics of the stock S of the products put out, but not yet consumed. The quantity S must obviously assume the form

$$\dot{S} = Y - C - \dot{K} \tag{21.5}$$

and obey the constraint $S \geqslant 0$. In this case, we need to alter the right sides of the system (21.1) because now it follows from (21.1) and (21.5) that $Y + \lambda S = \text{const}$ and the unlimited growth of Y is impossible. A new corrected and complemented model yet remains to be a toy, although it is more complex than the available one. We do not suppose that it makes sense to complicate the model in this way, and so the model will retain equations of the form (21.1) to which we add only the constraint on the consumption $c \leqslant Y$. We have to put up with a possible appearance of negative volumes of stocks.

21.5. Economic Growth and Statement of the Problem

We assume that Y grows with time and that the faster the growth of Y, the better. It follows from equations (21.1) and the constraint (21.4) that the fastest progress of the economy occurs in the case of the full employment and is defined by the trajectories

$$Y_*(t) = \left(\frac{\alpha}{\rho + l + b}\right)^{b-1} \frac{L_0}{B} e^{(\rho + l)t}$$

$$K_*(t) = \left(\frac{\alpha}{\rho + l + b}\right)^{b} \frac{L_0}{B} e^{(\rho + l)t} \tag{21.6}$$

$$C_*(t) = \left(\frac{\alpha}{\rho + l + b}\right)^{b-1} \frac{L_0}{B} \left(\frac{\rho + l}{\lambda} + 1 - \alpha + \frac{\alpha\beta}{\rho + l + \beta}\right) e^{(\rho + l)t}$$

We set

$$0 < \frac{\rho + l}{\lambda} + 1 - \alpha + \frac{\alpha\beta}{\rho + l + \beta} < 1,$$

in which case the constraint on the consumption is valid. The exponents (21.6) are spoken of as the trajectories of equilibrium growth. We will examine a problem for obtaining these trajectories in the system (21.1) since it is impossible to get any better result at all. If we apply to the system (21.1) the control action

$$C = ZY \tag{21.7}$$

with the coefficient

$$z = \frac{\rho + l}{\lambda} + 1 - \alpha + \frac{\alpha\beta}{\rho + l + \beta},$$

the variables Y and K and certainly C will move along the trajectories (21.6), but only if their initial values belong to this trajectories. But what happens under other initial conditions? Let us take it that

$$y = (Y - Y_*)e^{-(\rho+l)t}$$
$$k = (K - K_*)e - (\rho + l)t \qquad (21.8)$$

Values of y and k will be used to measure the variations of the real Y and K from the equilibrium trajectories. Note that the difference between two variables, for example, Y and Y^*, is commonly found from their absolute deviation; in this case, from a value of $Y - Y^*$. But we will use relative deviations (accurate to the scale factor), for example,

$$y = \frac{Y - Y_*}{Y_*}\left(\frac{\alpha}{\rho + l + \beta}\right)^{b-1}\frac{L_0}{B}$$

Here, we can readily cite examples of the functions $Y(t)$ for which $y \to 0$, but $|Y - Y_*| \to \infty$ when $t \to \infty$. Thus, at a low value of y, values of Y and Y^* differ from each other by a small percentage, rather than by a small number of units of measurement (dollars as of 1989). For the consumption (21.7), values of k and y obey the equations

$$\dot{K} = \alpha y - (\rho + l + \beta)K$$
$$\dot{y} = \frac{\lambda\alpha\beta}{\rho + l + \beta}y - \lambda\beta K \qquad (21.9)$$

The roots of the characteristic polynomial of this system are equal to zero and to

$$\frac{\lambda\alpha\beta}{\rho + l + \beta} - (\rho + l + \beta),$$

and so when $(\rho + l + \beta)^2 < \lambda\alpha\beta$, some of the solutions of the system extend to infinity. This means that the trajectories of equilibrium growth are unstable.

Let us set the objective so as to ensure the stability of equilibrium growth in the "large". In other words, we will try to replace (21.7) by the control law at which values of y and k specified by (21.8) approach zero from any initial point.

21.6. Linear Control and the Uncertainty Problem

The system under consideration is described by linear equations, and so it can be assumed that the linear control will be effective for the system. Let

$$w = (c - c_*)e^{-(\rho+l)t} \qquad (21.10)$$

The condition (21.7) means that $w = zy$, which, as became evident, does not offer us anything. Therefore, we discard (21.7). In the general case, the behavior of k and y obeys the equations

$$
\begin{aligned}
\dot{k} &= \lambda y - (\rho + l + \beta)k \\
\dot{y} &= \big(\lambda(\lambda - 1) - (\rho + l)\big) y - \lambda\beta k + \lambda\omega
\end{aligned}
\qquad (21.11)
$$

It is obvious that the control $w = m_1 k + m_2 y$ at certain values of m_1 and m_2 will ensure the motion of $k \to 0$ and $y \to 0$ in the system (21.11) under any initial conditions; the search for appropriate m_1 and m_2 is left to the reader. In view of (21.8), (21.10), which is alternative to (21.7), the policy of consumption is found to be the following:

$$c = c_* + m_1(k - k_*) + m_2(Y - Y_*) \qquad (21.12)$$

We could now finish the discussion if it were not for a certain circumstance. To use in practice the condition (21.12), it is necessary to know the current values of C^*, K^*, and Y^*. For this, we need the information on the parameters λ, β, ρ, b, B, l, and l_0. In all likelihood, the values of these parameters are not known exactly. Therefore, we cannot shape up the control (21.12).

Let us note a peculiar feature of the situation: the uncertainty of the parameters rules out not only the solution of the problem (21.12) for the tracking of K^* and Y^* by K and Y, but also the statement itself of this problem. It goes without saying that there is no way of tracking something that is not observable.

21.7. Control of Unemployment as Induction

To try to end the deadlock, it seems likely to ensure the condition of the full employment. If $L = L_*$ (both quantities are directly measurable), then

$$Y = \left(\frac{L_0}{B}\right)^{\frac{1}{b}} e^{\frac{\rho+l}{b}} k^{1-\frac{1}{b}} \qquad (21.13)$$

and from the first equation of (21.1) it follows that

$$\frac{\partial}{\partial t}(ke^{\beta t})^{\frac{1}{b}} = \frac{\alpha}{b}\left(\frac{L_0}{B}\right)^{\frac{1}{b}} e^{\frac{\rho+l+\beta}{b}t}$$

for which reason, $k \to 0$ and $y \to 0$ under any initial conditions.

Note that the condition (21.13) is nothing but the condition describing internal feedback that secures the desured character of motions in the system (21.1). The induction of this feedback represents control over the unemployment.

If there is a way to achieve an almost full employment, i. e., to secure the condition

$$1 - \frac{L}{L_*} \leqslant \varepsilon \qquad (21.14)$$

at any small value of $\varepsilon > 0$ (say, 2 or 3%), this points to the induction of internal feedback (21.13) with an implicitly prescribed error. From (21.14) it follows that

$$Y \geqslant \left(\frac{L_0(1 - \varepsilon)}{B}\right)^{\frac{1}{b}} e^{\frac{\rho+l}{b} t} k^{1-\frac{1}{b}}$$

Therefore, in view of the first equation of (21.1), we have

$$\frac{\partial}{\partial t}\left(k e^{\beta t}\right)^{\frac{1}{b}} \geqslant \frac{\alpha}{b}\left(\frac{L_0(1 - \varepsilon)}{B}\right)^{\frac{1}{b}} e^{\frac{\rho+l+\beta}{b} t}$$

and after a certain instant of time,

$$k e^{-(\rho+l)t} \geqslant (1 + \varepsilon)^{\frac{1}{b}}(1 - -2\varepsilon)^{1-\frac{1}{b}} \frac{L_0}{B}\left(\frac{\alpha}{\rho+l+\beta}\right)^{b-1}$$

from which we obtain

$$Y e^{-(\rho+l)t} \geqslant (1 + \varepsilon)^{\frac{1}{b}}(1 - -2\varepsilon)^{1-\frac{1}{b}} \frac{L_0}{B}\left(\frac{\alpha}{\rho+l+\beta}\right)^{b-1}$$

It is easy to select a value of $\varepsilon > 0$ such that the values $|y|$, $|k|$ become smaller in a finite time than any number prescribed beforehand. Thus, inequality (21.14) can secure a satisfactory character of motions.

21.8. Induction Control

We will solve the problem for the induction of feedback (21.13) with a certain error, the problem being stated implicitly in the form of the condition (21.14).

We did not examine the problems stated in this way in the main portion of the course book, so that we are devoid of the formula for the induction control. We need to derive completely the control law. In the case at hand, it can be done quite readily.

Let $\xi = \ln(L/L_*)$. In view of (21.1), (21.2), and (21.3), we have

$$\frac{\partial}{\partial t} \xi = (1 - b)\lambda \frac{Y}{K} - b\lambda\beta \frac{K}{Y} + b\lambda(\alpha - 1) - (1 - b)\beta - (\rho+l) + b\lambda \frac{C}{Y} \quad (21.15)$$

We now deal with the one-dimensional control problem known to us from Sec. 4. Its solution can be written as

$$C = YF(Y,K)\min\left\{a, \frac{L_* - L}{\varepsilon L_*}\right\} \tag{21.16}$$

where the function F satisfies the condition

$$F(Y,K) > \frac{1}{b\lambda}[(b-1)\lambda\frac{Y}{K} + b\lambda\beta\frac{K}{Y} + b\lambda(1-\alpha) - (b-1)\beta + (\rho + l)]$$

As an example of this function, we can take $F = A\big((Y/K) + (K/Y) + 1\big)$ at a rather large constant A. Obviously, on the strength of (21.15), the consumption (21.16) will force to increase steadily, at least until the condition $\xi \geqslant \ln(1 - \varepsilon)$ becomes steady. It should be noted that the use of the detailed information on the parameters of equation (21.15) enables us to obtain a less rough estimate of the right side of this control as against the function $A\big((Y/K) + (K/Y) + 1\big)$. The use of this estimate in the control law is justifiable on account of the constraint $C \leqslant Y$. We will not seek this estimate, the more so as it does not fully suit us. Indeed, under some initial conditions, the quantity Y/K or K/Y will be so high for certain that the control $C \leqslant Y$ turns out to be insufficient to effect the monotonic growth of ξ.

Moreover, there exist initial conditions under which any policy of consumption involving the constraint $C \leqslant Y$ is unable to afford the economic growth in the system (21.1). It is easy to verify this fact by considering a set of points $Y \leqslant (\beta/\alpha)K$. On the strength of the first equation of (21.1), here $\dot{K} \leqslant 0$, and in view of the second equation and the constraint on the control, we have

$$\dot{Y} - \frac{\beta}{\alpha}\dot{K} \leqslant \alpha\dot{K} - \frac{\beta}{\alpha}\dot{K} = 0$$

when $Y = (\beta/\alpha)K$. For this reason, the system cannot leave the domain in question where K does not increase and Y diminishes.

21.9. Comments

Since it is impossible to ensure the stability of the trajectories (21.6) in the "large", we did not achieve the assigned objective. Moreover, it turned out that if the state of the national economy is poor, then it will become much worse before long. In this situation, a real government evidently obtains credits abroad and does not adhere to the constraint $C < Y$ which we strictly observe. Besides, the real government does not bind itself to account for equations (21.1).

As regards the toy-type economy considered above, we were able to reveal the following.

1. The policy of the consumption (21.7) ensures equilibrium of the trajectories of growth, but does not afford their stability.

2. The policy of the consumption (21.12) ensures a local stability of these trajectories, but is unsuitable under the conditions of uncertainty of the parameters.

3. The policy of the consumption (21.16) achieves the same objective under the uncertainty conditions.

It is encouraging to note that in the case under study, the problem for the internal feedback induction coincides with the expert problem of control over unemployment. However, the fact that the full employment appears to be the sufficient condition of prosperity merely attests that the model (21.1) is unreal.

22. Technology:
Control of an Exothermic Reaction

In metallurgy, chemical industry, heat power industry, and in many other spheres of production engineering (including cookery), exothermic chemical reactions occur, i. e., reactions that proceed with the evolution of heat. The reader can discover an example of such a reaction in his kitchen when using the heat released in the course of the reaction between methane and oxygen. In this section, we examine problems for control of exothermic reactions involving two reage.

22.1. Production Process and the Control Problem

We will consider an industrial installation, such as a chemical reactor, a kiln, or any other unit or setup, in which reactions take place between certain agents. We assume that they arrive at the reaction space as components of the incoming flows in which the reagents are mixed with ballast substances, i. e., inactive agents that do not enter into the reaction. For example, a steam-engine firebox receives oxygen mixed with nitrogen and other gases (free air) and also carbon mixed with ash (coal).

Suppose that the installation of interest has a fixed volume and the reaction proceeds at a constant pressure. Then, the reaction space has a limited holding capacity. After its filling, which implies that the installation turns on to the operating conditions, the incoming flows of substances tend to form an outgoing flow. The substances that escape from the installation include chemical reaction products, ballast substances, and also remainders of the reagents that have not had time to react.

Assume that the reaction of interest occurs only at a fairly high temperature and the reagents get hot on exposure to the heat released during the reaction. Certainly, it is first necessary to initiate the reaction, for example, to strike a "match", so that the substances must catch fire and burn on their own. We can influence the course of the process after inducing the reaction by varying the rate of delivery of the reagents to the reaction space.

The assumptions made above are sufficient to describe qualitatively what effect the reagent flows of the appropriate rates exert on the course of the reaction. If the reaction space receives a large amount of reagents per unit time, each portion of the substance resides in the installation for a short time, and so an appreciable share of reagents has no time ro react and remains in the outgoing flow. Besides, the incoming flows heavily cool down the reaction space, with the result that the reaction can come to a halt. But if the rate of delivery of reagents is low and they have time to react almost completely, the reaction space is filled up both with "nonburning substances" (ballast) in the incoming flows and with the reaction products, in which case the reaction can also terminate.

Thus, we have clarified the concept of the control problem: there is a need to deliver reagents to the installation so as to sustain the continuous reaction.

22.2. Mathematical Model of the Process

As in the case of the nuclear reactor, here we will use a lumped model for describing a chemical reactor (Fig. 22.1). Assume that the reaction involves two reagents. At all points of the reaction space and in the outgoing flow, the mass concentrations of these substances are taken to be the same everywhere and equal to C_1 and C_2. The temperature in the reactor is also taken to be identical everywhere and equal to T. In this way, we preclude any nonhomogeneity in the reactor space, which points to the idealized lumped model. We suppose that the state of the process depends on the variables C_1, C_2, and T which obey equations of the form

$$\dot{C}_1 = \lambda_1 - \beta_1 k(T) C_1 C_2 - \gamma_1 C_1$$
$$\dot{C}_2 = \lambda_2 - \beta_2 k(T) C_1 C_2 - \gamma_2 C_2 \tag{22.1}$$
$$\dot{T} = \lambda_3 + \beta_3 k(T) C_1 C_2 - \gamma_3 T$$

where $k(T) = \exp(-\mu/T)$, $\lambda_i, \beta_i, \gamma_i$, and μ are positive parameters.

Let us elucidate the meaning of equations (22.1). First, we note that all the three equations of the model in the right sides contain summands proportional to $k(T)C_1 C_2$. This quantity corresponds to the process rate, i. e., to the number of individual acts of the chemical reaction, which occur per unit time in a unit mass contained in the reactor. We will not discuss why the process rate is preset by just

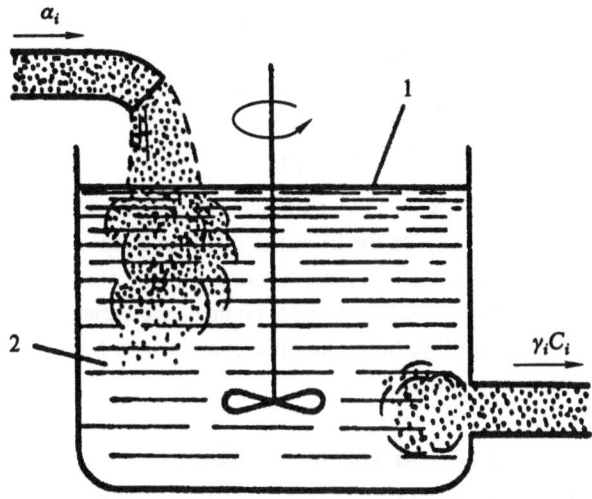

Fig. 22.1. The fluid fed into the chemical reactor loses its specific features and the concentrations of reagents become identical in each elementary volume. The concentrations of reagents C_1 and C_2 in the reactor and in the outgoing flow are the same.

this function, considering that the physicochemical science suffices to confirm our view.

The meaning of the first two equations of (22.1) is easy to understand if we consider that they describe changes in the number of molecules of the two reagents in the reaction space. In our case, these variables are proportional to the concentrations of the substances. Molecules of the ith reagent ($i = 1, 2$) arrive at the setup from the outside by means of the incoming flows (the summand λ_i), enter into the chemical reaction (the summand $-\beta_i k(T)C_1 C_2$), and move outside by way of the outgoing flow (the summand $-\gamma_i C_i$).

The third equation of (22.1) can be viewed as the equation that describes changes in the amount of heat energy in the reactor. The inflow of energy from the outside depends on the incoming flows (summand λ_3) and the outflow of energy depends on the outgoing flow and heat exchange with the environment (summand $-\gamma_3 T$). Besides, heat is evolved inside the reactor due to chemical interactions (summand $\beta_3 k(T)C_1 C_2$).

The input variables of the dynamic system under study show up in values of the parameters of equations (22.1). The rate of delivery of the ith reagent can be changed by varying the incoming flow that carries this reagent or by varying the concentration of the reagent in this flow. The magnitude of the incoming flow also has an effect on the outgoing flow, for which reason changes in the ith incoming flow must entail changes in the parameters λ_i, γ_i, α_3, and γ_3. At the same time, we can acknowledge to some extent that a change in the input concentration of the ith reagent will affect only α_i. So, when using this concentration as a control, the controllable object model will belong to the class of equation systems for which the course book includes the technique for developing control algorithms. In this connection, we assume that

$$\alpha_1 = \eta u \tag{22.2}$$

where u is the control. Changes in the parameter η, as well as changes in other parameters of equations (22.1), result from the effect of disturbances.

Frankly speeaking, practice relies on a more popular control that involves changes in the incoming flows. Clearly, these changes are easier to implement in practice. The control "lever" we have chosen is less feasible, but then it fits more closely the content of our course book.

22.3. Statement of the Control Problem

We will work out an algorithm of the control $u = u(C_1, C_2, T)$ so that it could ensure a state of the system (22.1), (22.2), which is close in a certain sense to the prescribed point C_1^*, C_2^*, T^*. The coordinates of this point cannot obviously be chosen arbitrarily. Really, if certain means can afford the equality $C_2 = C_2^*$ at $C_2^* = 0$, then from the third equation of (22.1) it follows that $T < 0$ when $T > \lambda_3/\gamma_3$, for which reason T cannot be too high. When the parameters of (22.1), (22.2) are constant, the prescribed state at a certain control $u = u^* = \text{const}$ can be thought of as a singular point of the system:

$$\eta u^* - \beta_1 k(T^*)C_1^* C_2^* - \gamma_1 C_1^* = 0$$
$$\alpha_2 - \beta_2 k(T^*)C_1^* C_2^* - \gamma_2 C_2^* = 0 \tag{22.3}$$
$$\alpha_3 + \beta_3 k(T^*)C_1^* C_2^* - \gamma_3 T_* = 0$$

In the general case, it is expedient to assume that equalities (22.3) are valid for the prescribed point in a certain "mean", "typical" set of values of the parameters or, at least, in one of the physically possible sets of these values. Then, the coordinates of C_1^* and C_2^* correspond to a nonsingular point of the system, so that one can only hope that the control will force the system to move near this point.

It is of interest to clarify what "near the assigned neighborhood" means. Is it possible, for example, to limit the motion of the system within the arbitrarily small preassigned neighborhood of the point C_1^*, C_2^*, T^*? Unfortunately, this cannot generally be done. The thing is that at some values of the parameters, other than those which are valid for equalities (22.3), the right side of the second or the third equation of (22.1) may prove to be alternating everywhere in the small neighborhood of the prescribed point. Then, the system will leave for sure this neighborhood. Note that we discussed the features of this kind in Sec. 16 (Example (16.7)) and in Sec. 19 for biological systems.

In our case, an achievable accuracy of the stabilization depends on the character of variations in the parameters α_2, α_3, β_2, β_3, γ_2 and γ_3. We will not consider this problem in detail.

Note that the problem of moving a system closer to the assigned point is not amenable to the solution under all the initial conditions, even under the conditions belonging to those which have a physical meaning. Clearly, the three-dimensional system (22.1) can move only in the domain $0 \leqslant C_1 \leqslant 1, 0 \leqslant C_2 \leqslant 1$, $0 \leqslant T \leqslant +\infty$. But in this case, it also makes sense to introduce additional constraints on T: at too low temperatures, the chemical reaction ceases, but at too high temperatures, the effects unaccounted for in the model make themselves evident, for example, the effect of the failure of the overheated setup. Assume that there is a range of temperatures $[T_{\min}, T_{\max}]$ beyond which the system should not go out and the assigned point T^* and the initial point $T(0)$ lie within this range rather far away from its bounds.

Thus, evolutions of the system at hand in the state space are bounded by a parallelepiped. This permits us to replace the nonlinear summands in the right sides of (22.1) by the linear function with varying limited coefficients. The equations derived in this case will appear simpler, and so there is a hope that they will allow us to use the results of the theory of linear control systems. But we will not embark on this path of simpliifications. The thing is that in passing to the linear notations, all the real nonlinearity will show up in the varying coefficients. This nonlinearity will be lost when these coefficients are found to be indeterminate. Thus, the linearization involves the loss of the information contained in the nonlinear notation. We cannot afford the careless treatment of the available information, for we are then likely to pay off in the value of the control action that represents the concentration of one of the reagents and obeys the inequality $0 \leqslant u \leqslant 1$. Therefore, later on we will deal with the available nonlinearity, i. e., with the system (22.1), (22.2).

22.4. Statement of the Induction Problem

A control has the direct effect on a change in the quantity C_1 and allows us to subject this coordinate to certain special conditions. In particular, it is possible to relate the current value of C_1 to the current values of C_2 and T, i. e., to induce internal feedback

$$C_1 = v(C_2, T) \tag{22.4}$$

What kind of feedback could impart the desired motions to the system? We substitute (22.2) into the second and third equations of (22.1) and denote

$W = K(T)C_2v$, for convenience. We obtain

$$\dot{C}_2 = \lambda_2 - \beta_2 w - \gamma_2 C_2$$
$$T = \lambda_3 + \beta_3 - \gamma_3 T \qquad (22.5)$$

It is evident that at $W = \text{const}$ (and at constant parameters), the system (22.5) has the only singular point (stable node) and the coordinates C_2 and T tend to appropriate values with the asymptotics $e^{-\gamma_2 t}$ and $e^{-\gamma_3 t}$. If

$$W = W^* = k(T^*)C_1^* C_2^*$$

then $C_2 \rightarrow C_2^*$ and $T \rightarrow T^*$, from which it follows that $C_1 \rightarrow C_1^*$.

Of course, this character of motions in the system occurs only when values of the parameters of equations (22.5) and (22.3) are coincident. In the general case, the pair of the variables C_2 and T will tend to the point with the coordinates $(\lambda_2 - \beta_2 W^*)/\gamma^2$ and $(\lambda_3 + \beta_3 W^*)/\gamma^3$, which obviously moves within a certain domain on account of changes in values of the parameters. The assignment (C_2^*, T^*) also lies within this domain, but it is possible to secure the tendency of only the point (C_2, T) toward a domain the size of which depends on the range of changes in the parameters.

We can increase the accuracy of stabilization of the system along the coordinates C_2 and T, i.e., reduce the above domain, through the use of another function W. We take the constant W^* used above as the basic value of W. Let $W = W^* + w$. Let us also determine $x_2 = C_2 - C_2^*$ and $x_3 = T - T_*$ and write equation (22.5) in the form

$$\dot{x}_2 = (\lambda_2 - \beta_2 W^* - \gamma_2 C_2^*) - \beta_2 w - \gamma_2 x_2$$
$$\dot{x}_3 = (\lambda_3 - \beta_3 W^* - \gamma_3 T^*) + \beta_3 w - \gamma_3 x_3 \qquad (22.6)$$

We thus have the linear system with the control w for which the law $w = k_2 x_2 - k_3 x_3$ is suitable. The reader can have the opportunity to verify on his own that at $k_2 \geqslant 0$ and $k_3 \geqslant 0$ and at constant parameters λ_i, β_i and γ_i, this system has a singular point of the stable node type. In the case of varying parameters, the point (x_2, x_3) tends to a certain neighborhood of the coordinate origin. It is worth noting that this neighborhood can be made as small as desired through an increase in k_2 and k_3, thus improving at the same time the asymptotics of the convergence of the system to the assignment. Yet, one should keep in mind that we deal with two out of the three coordinates of the object (22.1) and the behavior of C_1 does not improve in view of this fact; those who are in doubt can consider the example (16.7) once again and compare it with the present result.

Let us take it that those motions of the system (22.5) are satisfactory which correspond to the expression

$$W = W^* + k_2(C_2 - C_2^*) - k_3(T - T^*)$$

at certain values of $k_2 \geqslant 0$ and $k_3 \geqslant 0$. This equality indicates that the system (22.1) induces internal feedback of the form (22.4) with the operator

$$v = \frac{W^* + k_2(C_2 - C_2^*) - k_3(T - T^*)}{k(T)C_2} \tag{22.7}$$

However, up to now we have considered the system in which feedback (22.4), (22.7) is induced exactly, i. e., with the zero error. The continuous control cannot afford this.

Assume that internal feedback (22.4), (22.7) is induced with the error $\sigma = $ const > 0. In this case,

$$C_1 - v(C_2, T) = v\sigma \tag{22.8}$$

where $v = v(t) \in [-1, 1]$. Note that $C_2 \leqslant 1$ and $k(T) < 1$, so that

$$v + v\sigma = k^{-1}(T)C_2^{-1}(W^* + \varepsilon\sigma + k_2(C_2 - C_2^*) - k_3(T - T^*)),$$

where $|\varepsilon| \leqslant 1$. Thus, under the condition (22.8), motions of the system (22.5) assume the form

$$W = W^* + \varepsilon\sigma + k_2(C_2 - C_2^*) - k_3(T - T^*)$$

which differs from the expression considered above only by the free term $\varepsilon\sigma$. Obviously, at a fairly small σ, the effect of this summand is low. Suppose that the error σ is chosen with due regard for this concept.

The operator $v(C_2, T)$ specified by (22.7) and the induction error σ determine the statement of the induction problem.

22.5. Synthesis of the Induction Control

Let $u = u^* + \bar{u}$, where u^* is the basic value of the control, which satisfies equality (22.3) at a certain set of the parameters. As always, we denote $s = C_1 - v(C_2, T)$ and, in view of (22.7), obtain

$$\dot{S} = \dot{C_1} - \frac{\partial v}{\partial C_2}\dot{C_2} - \frac{\partial v}{\partial T}\dot{T}$$

$$= \eta\bar{u} + \eta u^* - \beta_1 k(T)C_1 C_2 - \gamma_1 C_1 -$$

$$+ \frac{k_2 C_2^* + k_3(T - T^*) - W^*}{k(T)C_2^2}(\alpha_2 - \beta_2 k(T)C_1 C_2 - \gamma_2 C_2) +$$

$$+ \frac{k_3 + \dfrac{M}{T^2}\left[W^* + k_2(C_2 - C_2^*) - k_3(T - T^*)\right]}{k(T)C_2}(\lambda_3 + \beta_3 k(T)C_1 C_2 - \gamma_3 T)$$

or, introducing for convenience the requisite designation, we have

$$\dot{s} = \eta\bar{u} + f(C_1, C_2, T, t) \qquad (22.9)$$

where the dependence of f on time points to the fact that the parameters α_i, β_i, γ_i can vary. Let

$$h > 1/\inf \eta, \quad F(C_1, C_2, T) \geqslant |f(C_1, C_2, T, t)|$$

According to the general technique, the induction control $\bar{u}$ can be written as

$$\bar{u} = -hF(C_1, C_2, T) \min\left\{1, \frac{|s|}{\sigma}\right\} \operatorname{sgn} s \qquad (22.10)$$

22.6. Constraints and Drawbacks

The control law (22.10) includes the majorant of the absolute value of the cumbersome function f which need be estimated quite accurately so as not to violate the constraint $-u^* \leqslant \bar{u} \leqslant 1 - u^*$. There is a way of obviating this tedious calculation if we include the constraint itself into the control law. To clarify this suggestion, we note that formula (22.10) describes a smoothed change-over of the control $\bar{u}$ from a value of $+hF$ to a value of $-hF$. Instead of this change-over, we can resort to the switching of u form 1 to 0:

$$u = \begin{cases} 1 & \text{for } s < -\sigma \\ s/\sigma & \text{for } |s| \leqslant \sigma \\ 0 & \text{for } s > \sigma \end{cases} \qquad (22.11)$$

The control law (22.11) is set up in a much simpler way than (22.10), but the former does not solve the induction problem with the same success. Formula (22.10) has its advantages: if the majorization is carried out rather accurately, then the values of $+hF$ and $-hF$ can turn out to be closer to each other than 0 and 1 at certain values of the coordinates. Here, the control (22.10) has a lower gradient than (22.11); this is of importance for practice.

However, at some points of the state space the smallest of the majorants, too, can take on values at which the control (22.10) cannot fall within the constraints. Consequently, there is no way of ensuring a decrease in $|s|$ at these points. It is not inconceivable that under some initial conditions, it is impossible to induce internal feedback (22.4), (22.7) by a control that has a physical meaning. Moreover, even a meaningless unbounded control cannot sometimes do anything: if a value of the function v defined by (22.7) falls outside the section [0, 1], equality (22.11) makes no sense and there is no way of deriving it for the system.

The above reasoning suggests the idea of examining the following.

1. Under what conditions does the constructed induction system operate?

2. In what cases does the technique inplemented in a certain way for the induction of internal feedback loops become in principle effective?

3. When is the stated control problem resolvable at all?

The answers to these questions cannot be found without introducing additional suppositions as to the behavior of the system parametrers. If, for example, the quantity λ_2 goes to zero and remains in this state for a long time, then the coordinate C_2 will begin to approach zero irrespective of the control, which causes a decrease in temperature, so that the reaction can cease. Unfortunately, we should acknowledge that the comprehensive analysis of these effects would require considerable labor efforts on the part of the authors and readers. Therefore, we will not go to the heart of the matter and assume that it is more convenient to perform the analysis of the above problems individually in each specific practical case when there is a large body of information on the properties of the system. Here, we content ourselves with the fact that there exist initial conditions and parameters at which the developed algorithms behave in the conceived way.

23. Medicine: Control of Carbohydrate Metabolism for Diabetes

Hormones take part in many processes occurring in a human organism. An acceptable character of the occurrence of these processes depends on the functioning of numerous feedback loops granted to a human being from his birth; the hormones play the role of control signals in these loops. If the mechanism of the production (secretion) of any hormone breaks down, the behavior of the organism as a dynamic system may take undesirable features even to the extent that the human life can be in danger. In this case, it is necessary to inject an appropriate hormone from the outside, thus replacing the natural feedback loop by the artificial one. In this section, we consider the algorithm of operation of this kind of loop.

23.1. Carbohydrate Metabolism and Insulin-Dependent Diabetes

Glucose is the basic energy carrier in the organism. In comes from food to the blood which delivers it to all the organs and tissues to supply them with the chemical energy stored in glucose. In the blood of a healthy human being, the concentration of glucose varies from 80 to 100 mg% (i. e. 1 mg in 100 ml of blood) before a meal and 140—180 mg% after a meal. The concentration falls outside this range on account of a more or less grave unhealthy state due to the hyperglycemia resulting from an excess of glucose and due to the hypoglycemia involving a shortage of glucose.

The concentration of glucose in the blood mainly depends on the following factors: diet, secretion of two hormones of the pancreas, namely, insulin and glucagon, and also glucosuria (only in diabetic patients), i.e., the ijection of glucose from the organism through the kindneys. We will describe qualitatively the role of these factors.

The delivery of glucose at a certain time to the blood from the digestive system obviously depends on how much food, what kind of food, and when a human being ate it.

As the amount of glucose increases, the pancreas secretes insulin that gives impetus to the lever and also to muscular and adipose tissues so that they store (deposit) glucose by recovering it from the blood. As a result, the amount of glucose in the blood is bounded above.

As the amount of glucose decreases, the secretion of a hormone-glucagon occurs, with the result that the stored glucose returns to the blood, thus limiting a decrease in the glucose concentration.

Diabetes disturbs the above processes of the carbohydrate metabolism. Diabetes of the widespread form (insulin-dependent diabetes) entails an insufficient secretion of insulin or the complete lack of its secretion. The glucose concentration then rises without any control despite the fact that some amount of sugar begins to flow out from the blood to urine through the kindneys. If the amount of glucose reaches 500 mg%, the comatose condition can result, which is fraught with the fatal outcome.

It should be said that insulin-independent diabetes also met with, in which case the lever affected by cirrhosis fails to respond to insulin and does not store glucose. We will not discuss this form of disease.

A shortage of insulin in the case of diabetes can be offset by delivering it from the outside in individual portions (injection) or in a continuous way (infusion). Of much importance here is a corectly estimated dose of the hormones introduced: an excess of insulin may lead to the hyperglycemia which is as troublesome as the hypoglycemia resulting from the insulin shortage. The search for an acceptable regimen of introducing insulin represents the essence of the control problem that we will deal with below.

23.2. Mathematical Model of Carbohydrate Metabolism

We take it that carbohydrate metabolism processes conform to the behavior of the second-order dynamic system with the state coordinate G (the concentration of glucose in blood) and the state coordinate I (the concentration of insulin). Assume that an increase in G results from the fact the a patient takes food and the lever produces glucose, but a decrease in G stems from both the deposition of glucose under the effect of insulin and the removal of glucose from the organism together with urine: the latter effect arises as the quantity G reaches a definite threshold value. Suppose that I increases only due to the artificial injection (the secretion of insulin in the organism is absent), but increases due to the decomposition (known as clearance) of insulin in the lever and peripheral tissues.

Note that the assumption as to the two-dimensionality of the system represents the idealization of the lumped model, which is known to us from the examples presented above. Hence, various processes of the propogation and the circulation of the substance in the organism are left out of consideration. For this reason, even the most refined equations in the two-dimensional model cannot be adequate to reality. So, the reader should not overstate the significance of our further discussion of applied nature. This aspect will be dealt with in detail in Sec. 23.4.

We will describe carbohydrate metabolism processes by the equations

$$
\begin{aligned}
\dot{G} &= f_1 + f_2 - y - aGI \\
\dot{I} &= u - bI
\end{aligned}
\tag{23.1}
$$

The summands in the right side of the model have the following meaning.
1. f_1, the rate of delivery of glucose from the gastroenteric tract.
2. f_2, the rate of release of deposited glucose under the action of glucagon.
3. y, the rate of glucosuria.
4. aGT, the deposition rate of glucose.
5. u, the rate of delivery from the outside of insuline to the blood;
6. bT, the rate of clearance (decomposition) of insulin.

We will introduce a number of assumptions for values of the model parameters.

The quantity $f_1 = f_1(t)$ is equal to zero before a meal, steadily grows after a meal, and then drops to zero in a way similar to the slope of one arch (half-wave) of the sinusoid. The span of this arch depends on the menu: the oatmeal porridge produces the arch with a span of nearly 2.5 hours and pure sugar does so with a span of 1 hour. The height of the arch depends on the amount of food, more accurately, on the volume of glucose eaten.

The summand $f_2 = f_2(t)$ is equal to zero at relatively high values of G, but as G decreases, the summand increases to a certain limiting value. Generally speaking, a value of f_2 depends on a number of factors which we disregard; we restrict ourselves to the statement of the fact that $f_2(t)$ is a bounded nonnegative function.

As noted above, the effect of the glucosuria appears when the quantity G attains a certain threshold value of G^*. We can assume that

$$y = y(G) = \max\{0, c(G - G^*)\} \tag{23.2}$$

The quantity u will be taken as the input of the dynamic system. Other summands in the right side have no need for the clarification.

23.3. Constraints on Variables and Statement of the Problem

The desires of medical men in regard to the behavior of the system (23.1) can be given in the form: for three preassigned positive numbers $G_1 < G_2 < G_3$ there is a need to secure both the condition $G_1 \leqslant G \leqslant G_2$ on an empty stomach and the inequality $G_1 \leqslant G \leqslant G_3$ on a full stomach. We assume that the threshold of the glucosuria G lies between G_2 and G_3. The behavior of I is not subject to the regulation in any way.

To achieve the objective at hand, it is permissible to use the control $u \in [0, U]$, the constraints follow from the potential of the medical technology.

The stated objective is insolvable if we do not introduce additional constraints.

1. The clearance of insulin must proceed rather slowly (the number b is small) so that I can be raised markedly due to a limiteid value of u.

2. The liver must effectively respond to insulin (the number a is fairly high) so that G can be reduced due to an increase in I.

3. A diabetic cannot permit himself to eat at once too large an amount of carbohydartes (the number f_1 is small), otherwise no cure will be effective.

4. Before a meal, a sufficient quantity of glucose must enter into the blood from the lever (the number f_2 is not too small) so as to preclude the hypoglycemia. This constraint actually means that the patient must not be hungry too long.

Formally, the additional constraints can be given as the inequalities

$$\sup(f_1 + f_2) \leqslant \inf a \cdot G_3 \frac{U}{\sup b} + \inf C \cdot (G_3 - G^*),$$

$$\sup f_2 \leqslant \inf a \cdot G_2 \cdot \frac{U}{\sup b}$$

$$\inf(f_1 + f_2) > 0$$

Under these conditions, we will seek the function $u = u(G, I)$ that ensures the desired behavior of G in view of (23.1).

23.4. Is Diabetes Doomed?

An exclusively applied meaning of the problem of the cure for diabetes forces us to digress from the subject and discuss practical prospects for the inferred result.

Assume that the desired function $u = u(G, I)$ is found and the model (23.1) with the appropriate control behaves in the required way. We will also make an assumption that the model (23.1) satisfactorily defines the carbohydrate metabolism in an organism.

To use the desired theoretical results, we need to have the following: (1) measuring devices intended to measure the current values of $G(t)$ and $I(t)$; (2) computing devices intended to estimate the current values of the function $u(t) = u(G)(t)$, $I(t)$; and (3) actuators for the infusion of insulin at a varying rate $u(t)$. It is naive to believe that all these devices are ready for use in each hospital and stand waiting for our results.

We will not describe the harrowing state of native medicine, but turn our attention to the civilized world and to advanced specimens of medical equipment. It appears that all the devices are available in this world, excepting insulin analyzers. Hence, there is no way of measuring directly the current values of $I(t)$.

Is it possible to dispense with this measurement? Practice shows that it is possible. Apparatus of the artificial pancreas (stationary hospital units) obviate the need for the measurement of $G(t)$; the natural pancreas also seems to secrete doses of hormones without regard for the produced amount of insulin.

Most patients are given sufficient aid through the infusion of insulin doses estimated on the principles of the compensation for disturbances, diregarding feedback at all (the more food the patent eats, the larger the dose); the results of casual measurements of G serve to adjust the parameters used in the estimation of compensating actions. As a matter of fact, the healthy pancreas also uses information on disturbances, which it receives from what are known as gastrointestional hormones.

The results presented in our work disregard data on disturbances, on the one hand, and need data on insulin, on the other hand. This impracticability certainly causes concern. To justify ourselves, we will make three comments. First, data on disturbances (i. e., on food) only represent the prediction of values of $G(t)$, which is highly inaccurate: the content of carbohydrates of a common meal, excepting a special dietery food, is subject only to a rough estimation, and the process of transferring these carbohydrates to the blood under different conditions proceeds not identically. Second, insulin analyzers will be built up some time or other. Besides, if detailed studies are made of the process of the insulin clearance (the coefficient b in the second equation of the system (23.1)), it is possible to estimate, rather than to measure, the current values of $I(t)$. Third, the healthy pancreas does not rely on values of $I(t)$ and hence does not function in the best way. As some time passes after good dinner, a healthy human being often wants some addition of food, which is the symptom of the hypoglycemia resulting from the overproduction of insulin. Note that the natural way of avoiding the hazardous situation is to use an additional control (additional food), but in Sec. 23.3 we agreed to consider only u as a control.

One should not forget the basic aspect: in this section we deal with the mathematical problem of control of the dynamic system (23.1) rather than with the medical problem of control of diabetes. The statement of our problem involves a practical aspect in some way, but no more. Practical workers will judge whether the result will be appropriate in practice. We do not lay claim to play the role of saviors of mankind from the grave ailment.

23.5. Synthesis of the Control

We will construct the control $u(G, I)$ that will serve as a tool of inducing in the system (23.1) an internal feedback loop of the form $I = v(G)$ with the error σ, i.e., the tool of deriving the inequality

$$|I - v(G)| \leqslant \sigma \qquad (23.3)$$

Let $\sigma = \text{const} > 0$. In the space of the coordinates G and I, inequality (23.3) now prescribes a curvilinear band of width 2σ. We first clarify the values of v at which the fulfillment of (23.3) secures the desired behavior of the system and then think over how to obtain this inequality, namely, how to shift the point (G, I) to this band and hold it forever.

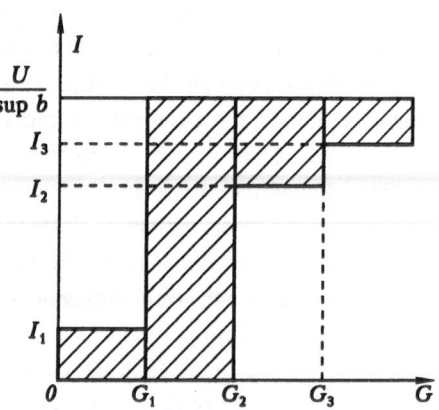

Fig. 23.1. The behavior of the system in the dashed region satisfies the requirements for the statement of the problem. However, it is impossible to make this region invariant, i.e., hold the system within it. There is a need to select in the dashed region a certain band that can be imparted the properties of invariance.

Note that in view of the first equation of (23.1), we have

$$\dot{G} \geqslant 0 \quad \text{for } G \leqslant G_1 \qquad\qquad \text{if } I \leqslant I_1 = \inf(f_1 + f_2)/G_1 \sup a$$

$$\dot{G} \leqslant 0 \quad \text{for } G_2 \leqslant G_1, f_1 = 0 \qquad \text{if } I \geqslant I_1 = \sup f_2/G_2 \inf a$$

$$\dot{G} \leqslant 0 \quad \text{for } G_3 \leqslant G \qquad\qquad \text{if } I \leqslant I_3 = \sup(f_1 + f_2)/G_3 \inf a$$

The above constraints on $\dot{G}$ at different values of G define the requirements for the system behavior, while the appropriate constraints on I represent means for satisfying these requirements. Assume that the condition $I \leqslant v/\sup b$ is compatible with all these inequalities. In this case, in the plane of the coordinates G and I, the band (23.3) must extend along the dashed region, as illustrated in Fig. 23.1. Let us suppose that $\sigma < 1/2$ and derive the function

$$v(G) = \begin{cases} \sigma & \text{for } G \leqslant G_1, \\ A_1 G + B_1 & \text{for } G_1 < G < G_2 \\ I_2 + \sigma & \text{for } G = G_2 \\ A_2 G + B_2 & \text{for } G_2 < G < G_3 \\ I_3 + \sigma & \text{for } G_3 \leqslant G. \end{cases} \tag{23.4}$$

The constants A_1, A_2, B_1, and B_2 are preset in (23.4) under the conditions of the continuous function v:

$$A_1 = \frac{I_2}{G_2 - G_1}, \qquad B_1 = \frac{G_2 \sigma - G_1(I_2 + \sigma)}{G_2 - G_1}$$

$$A_2 = \frac{I_3 - I_2}{G_3 - G_2}, \qquad B_2 = \frac{G_3(I_2 + \sigma) - G_2(I_3 + \sigma)}{G_3 - G_2}$$

An appropriate band defined by (23.3) appears in Fig. 23.2. The band is sited so that under any initial conditions, the point (G, I) enclosed in the band will shift to the inclined broken section after a meal and to the lower section of the region before a meal. This character of motions suits us.

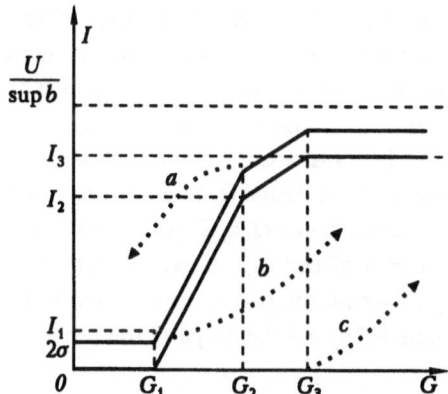

Fig. 23.2. The broken band can be made invariant. Undesirable trajectories of the types a and b can be omitted by introducing additional realistic constraints on the system parameters and trajectories of the type c can be ruled out by imposing constraints on the initial conditions.

We will now work out the control $u = u(G, I)$ that will set up the condition (23.3) in the system (23.1) after a lapse of some time. Let us introduce designate s

as $s = I - v(G)$ and determine

$$u(G, I) = U \max \left\{ 0, \min \left\{ 1, \frac{1}{2} \left(1 - \frac{s}{\sigma} \right) \right\} \right\} \qquad (23.5)$$

This function is equal to zero above the band (23.3), to v below the band, and takes intermediate values within the band.

23.6. Properties of the Closed System

Does the control (23.3) attend to its duty, namely, does it shift the point (G, I) to the band mentioned above and does it forbid the point to move outside? Generally speaking, this control does all that it can: below the band (23.3) it pulls the image point of the system upward at a maximum possible speed and allows it to fall freely from the top of the band. Nevertheless, this range of control is insufficient under certain conditions. Indeed if the slope of inclined sections of the band (coefficients A_1 and A_2) is relatively substantial and the ratio $\dot{I}/\dot{G}$ is small in the absolute value, the system may not restrain itself within the inclined sections. In this case, there may appear the hypoglycemia or hyperglycemia; the system behavior will not conform to the imposed requirements.

To do away with this situation, we need to impose a variety of new constraints: the numbers inf b, U are high, the number $\sup(f_1 + f_2)$ is fairly small, and the numbers G_1, G_2, and G_3 lie rather far away from one another. We omit the formal discussion of this aspect since it involves tedious and cumbersome efforts and only note that this task takes place in the case where specialists have to do with actual number ranges of the parameters of the model (23.1). After this stipulation, we can suppose that the control (23.5) handles the stated problem.

Let us point out that the control transfers the system (23.1) to the band (23.3) in the zero time. For this reason, at the begining of the control process, the system may behave in an undesirable manner. For example, if after a meal the initial point has the coordinates $G = G_3$ and $I = 0$, then the hyperglycemia is unavoidable and no control can save a patient from it. In this situation, it is evidently necessary to secure the patient in a certain way against these initial conditions, but this issue is beyond the scope of the problem considered above.

Conclusion

What we wish least of all is that the reader might evaluate our work by the hackneyed metaphor: "one more brick in the building of science". Such a building does nor exist, nor do the bricks. The brick is a solid, dead, and utilitarian piece, whereas scientific theories are living, movable phenomena of consciousness, which continuously penentrate with their roots and branches into one another and into the surrounding "unscientific" world. Science differs from creations of the human spirit not in the registration for living in a certain building, but in specific principles of the internal arrangement, much as trees differ by nature from other plants. We wish to think that our work is a fetus of such an "organism"?

The growth of this fetus depends on us and you, dear reader. What prospects can we look through here. We can single out two basic aspects of growth: the expansion to the "unscientific" field (i. e., the field of application) and purely theoretical development.

The trend for application studies involves the penetration into more and more new areas of practice, for which the theory is applicable. Mankind can move to this "global domination" by two routes. First, it is possible to take stock of the applied problems of control and varify those areas of application which are subject to the immediate annexation. In this case, one cannot cherish hopes for finding the collection of statements of the unsolved applied problems: the statements of the problems are made up so as to use definite means (methods) for the solution of these problems. Second, it is possible to rearrange in our interests some areas of application and make them convenient for taking hold other new areas. The matter is that an apprecial portion (to put it mildly) of controllable objects are artificial dynamic systems, including a variety of units for production processes. These systems can be designed so as to facilitate the use of the theory set forth in this work.

Theoretical investigations for the progression of the aspects described above can be undertaken in the most different directions, naturally without any hope for the success. It is of interest to consider, for example, the following.

1. How to extend the class of "elementary" dynamic systems that serve as intermediate units for the transfer of control actions.

2. How to shape up an induction control on the principle of feedback with respect to the output of an object if the output is not identical to the state.

3. How to implement the idea of the transfer of actions through intermediate units in the systems not describable by ordinary differential equations solved for derivatives.

The reader can put a variety of similar questions to himself, casting doubt sequentially on all assumptions and constraints which we explicitly or implicitly adopted in presenting the results.

Bibliography

Aizerman M. A. Automatic Control Theory. Third Edition (in Russian), Nauka, Moscow, 1966, p. 450.

Amemiya T. Stability analysis of nonlinearity interconnected systems — application of M-functions. Journal of Math. Analysis & Applic., 1986, No. 114, pp. 252–277.

Arnold V. I. Ordinary Differential Equations (in Russian). Nauka, Moscow, 1984, p. 272.

Astrom K. I., Borisson V., Ljung L., Wittenmark B. Theory and application of self-tunning regulators. Automatica, 1977, vol. 13, pp. 457–476.

Basar T. Minimax controllers for LTI plants under LTI-bounded disturbunces. 11-th IFAC World Congress. Tallinn, 1990, vol. 5. pp. 158–163.

Belforte G., Tay T. Optimal input design for worst-case system identification in $l^1 | l^2 | l^\infty$. Systems and Control Letters, 1993, vol. 20. pp. 273–278.

Bergstron A. The Construction and Use of Economic Models. English University Press, London, 1967.

Coppel W. A. Stability and asymptotic behavior of different equations. D. C. Heath, Boston, 1965.

Dahleh M. A. Bibo stability, robustness in the presence of coprime factors perturbations. IEEE Trans. Aut. Control. 1992, vol. AC-37, March, pp. 352–354.

Emel'yanov S. V. Automatic Control Systems of Variable Structures (in Russian), Nauka, Moscow, 1967, p. 336.

Emel'yanov S. V. Binary Systems of Automatic Control (in Russian). MNIIPU, Moscow, 1984, p. 313.

Emel'yanov S. V. (ed.). The Theory of Variable Structure Systems (in Russian), Nauka, Moscow, 1970, p. 592.

Emel'yanov S. V. and *Korovin S. K.* Use of the principle of error-closing control for the extension of sets of feedback loops, Dokl. Akad. Nauk SSSR, 258, No. 5, 1070–1074 (1981).

Emel'yanov S. V. and *Korovin S. K.* Extension of sets of feedback loops and their application in design of closed-loop dynamic systems, Izv. Akad. Nauk SSSR, Tekh. Kibern. No. 5, 173–183 (1981).

Emel'yanov S. V. and *Korovin S. K.* Principles of the design and basic properties of closed-loop dynamic systems with different types of feedback loops, in: Collection of Works "Dynamics of Inhomogeneous Systems", Proc. of Seminar, VNIISI, Moscow, 1982, pp. 5–27.

Emel'yanov S. V. and *Korovin S. K.* New types of feedback loops and their application to closed-loop dynamic systems. Results of Science and Engineering, Ser. Tekh. Kibern. VINITI, Moscow, 15, 145–216 (1982).

Emel'yanov S. V., Korovin S. K. and *Sizikov V. I.* Principles of the design and properties of control systems with integral coordinate-parametric feedback loops. Izv. Akad. Nauk SSSR, Tekh. Kibern., No. 6, 140–152 (1981).

Emel'yanov S. V., Korovin S. K. and *Sizikov S. I.* On the synthesis of nonlinear control of free motion of nonstationary systems, Dokl. Akad. Nauk SSSR, 265, No. 2, 297–302 (1982).

Emel'yanov S. V., Korovin S. K. and *Ulanov B. V.* On the synthesis of control systems with the use of coordinate-parametric and parametric feedback loops, Dokl. Akad. Nauk SSSR, 266, No. 5, 1077–1082 (1982).

Emel'yanov S. V., Korovin S. K. and *Ulanov B. V.* On the synthesis of control systems with quasicontinuous shaping of control actions, Dokl. Akad. Nauk SSSR, 268, No. 5, 1067–1071 (1983).

Emel'yanov S. V. and *Korovin S. K.* Control under Uncertainty Conditions:New Types of Feedback Loops (in Russian), Nauka, Moscow (1997).

Emel'yanov S. V., Korovin S. K. Development of feedback types and their application to design of closed-looped dynamical systems. Problems of Control and Information Theory. — Budapest, Publ. House of the Hung. Acad. of Sc., vol. 10(3), 1981, pp. 161–174.

Emel'yanov S. V., Korovin S. K., Sizikov V. I. Use of coordinate-parametric feedback in design of control systems. Problems of Control and Information Theory. — Budapest, Publ. House of the Hung. Acad. of Sc., vol. 10(4), 1981, pp. 237–251.

Emel'yanov S. V., Korovin S. K., Sizikov V. I. Control of non-stationary plants with coordinate-parametric feedbacks. Problems of Control and Information Theory. — Budapest, Publ. House of Hung. Acad. of Sc., vol. 11 (4), pp. 259–269, 1982.

Emel'yanov S. V. Bynary control systems, Mir Publ. House, 1984.

Emel'yanov S. V.,Korovin S. K.,Mamiconov I. G. Variable Structure Control Systems: Discrete and Digital. Mir Publ. House. — CRC, 1995.

Emel'enov S. V., Korovin S. K., Ulanov B. V. Control of nonstationary dynamic systems with quasicontinuous generation of the control signal. Problems of Control and Information Theory, vol. 12, No 1, 1983, pp. 11–32, Publ. House of Hung. Acad. of Sc., Budapest.

Fomin V. N., Fradkov A. L. and *Yakubovich V. A.* Adaptive Control of Dynamic Objects (in Russian), Nauka, Moscow, 1981, p. 448.

Francis B. A. A course in H^∞ control theory. Lect. Notes in Contr & Info Sci. 1987, vol. 88, New Iork, p. 156.

Furata K. Sliding mode control of a discrete systems. Systems & Control Letters, 1990, No. 14, pp. 145–152.

Goryachenko V. D. Qualitative Methods in the Dynamics of Nuclear Reactors (in Russian). Collection. Physics of Nuclear Reactors. Issue 23. Energoizdat, Moscow, 1983, p. 83.

Kailath T. Linear control systems. Prentice-Hall, Englewood-Cliffs, New Jersey, 1980.

Kurzhanskii A. B. Control and Observation under Uncertainty Conditions (in Russian), Nauka, Moscow, 1977, p. 392.

Lasalle I. P. The stability of Dynamical Systems. SIAM, Philadelphia, Pennsylvania, 1976.

Lee E. B., Markus L. Foundations of Optimal Control Theory. Wiley, New Iork, 1962.

Leondes C. T. (ed.) Decentralized-Distributed Control and Dynamic Systems. Part I. Acad. Press, New Iork, 1985.

Luenberger D. G. Canonical forms for linear multivariable systems. IEEE Trans. on Autom. Contr., 1967, vol. AC-12, pp. 290–293.

Mesarovic M. D. and *Takahara Y.* General System Theory. Mathematical Foundations. Academic Press, NY, 1974, p. 312.

Narendra K. S., Valavani L. S. Stable adaptive controller design — direct control. IEEE Transaction on Autom. Contr., vol. AC-23, pp. 570–583.

Novosel'tseva V. N. (ed.). Engineering Physiology and Simulation of Organism Systems (in Russian). Nauka, Novosibirsk, 1987, p. 236.

Parks P. C. Liapunov redesign of model reference adaptive control sytems. IEEE Transactions on Autom. Contr., 1966, vol. AC-11, pp. 362–367.

Perlmutter D. D. Stability of Chemical Reactors. Prentice-Hall, 1972.

Rudin W. Principles of Mathematical Analysis. McGraw-Hill, NY, 1964, p. 320.

Saberi A., Sannuti P. Global stabilization with almost disturbance decoupling of a class of uncertain nonlinear systems. International Journal of Control, vol. 47, pp. 717–727.

Safonof M. G. Future directions in H^∞ robust control theory. 11-th IFAC World Congress, 1990, Tallinn, pp. 147–151.

Scherer C. W. The state feedback H^∞ problem at optimality. Automatica, 1994, vol. 30, No 2, pp. 293–305.

Siljak D. D. Decentralized control of Complex Systems. Acad. Press, Inc., New Iork, 1991.

Tou Julius T. Modern Control Theory, NY, 1964, p. 471.

Tsypkin Ya. Z. Basics of Automatic System Theory (in Russian), Nauka, Moscow, 1977, p. 560.

Tsypkin Ya. Z. Adaptation and Learning in Automatic Systems (in Russian), Nauka, Moscow, 1968, p. 400.

Vidyasagar M. New directions of research in nonlinear control theory. Proceeding of the IEEE, 1986, No 74, pp. 1060–1091.

Vidyasagar M. Optimal rejection of persistent bounded disturbunces. IEEE Trans. on Autom. Control, 1986, vol. AC-31, June, pp. 527–534.

Volterra V. Theorie mathematique de la lutte pour la vie. Gauthier-Villars et C, Editeurs, Paris, 1931.

Voronov A. A. Basics of Automatic Control Theory (in Russian), Part 1, Energia, Moscow, 1965, Part 2, Energia, Moscow, 1966.

Wonham W. M. Linear Multivariable Control: A Geometric Approach. Springer, New Iork, 1979.

Lecture Notes in Control and Information Sciences

Edited by M. Thoma

1993–1998 Published Titles:

Vol. 186: Sreenath, N.
Systems Representation of Global Climate
Change Models. Foundation for a Systems
Science Approach.
288 pp. 1993 [3-540-19824-5]

Vol. 187: Morecki, A.; Bianchi, G.;
Jaworeck, K. (Eds)
RoManSy 9: Proceedings of the Ninth
CISM-IFToMM Symposium on Theory and
Practice of Robots and Manipulators.
476 pp. 1993 [3-540-19834-2]

Vol. 188: Naidu, D. Subbaram
Aeroassisted Orbital Transfer: Guidance
and Control Strategies
192 pp. 1993 [3-540-19819-9]

Vol. 189: Ilchmann, A.
Non-Identifier-Based High-Gain Adaptive
Control
220 pp. 1993 [3-540-19845-8]

Vol. 190: Chatila, R.; Hirzinger, G. (Eds)
Experimental Robotics II: The 2nd
International Symposium, Toulouse,
France, June 25-27 1991
580 pp. 1993 [3-540-19851-2]

Vol. 191: Blondel, V.
Simultaneous Stabilization of Linear
Systems
212 pp. 1993 [3-540-19862-8]

Vol. 192: Smith, R.S.; Dahleh, M. (Eds)
The Modeling of Uncertainty in Control
Systems
412 pp. 1993 [3-540-19870-9]

Vol. 193: Zinober, A.S.I. (Ed.)
Variable Structure and Lyapunov Control
428 pp. 1993 [3-540-19869-5]

Vol. 194: Cao, Xi-Ren
Realization Probabilities: The Dynamics of
Queuing Systems
336 pp. 1993 [3-540-19872-5]

Vol. 195: Liu, D.; Michel, A.N.
Dynamical Systems with Saturation
Nonlinearities: Analysis and Design
212 pp. 1994 [3-540-19888-1]

Vol. 196: Battilotti, S.
Noninteracting Control with Stability for
Nonlinear Systems
196 pp. 1994 [3-540-19891-1]

Vol. 197: Henry, J.; Yvon, J.P. (Eds)
System Modelling and Optimization
975 pp approx. 1994 [3-540-19893-8]

Vol. 198: Winter, H.; Nüßer, H.-G. (Eds)
Advanced Technologies for Air Traffic Flow
Management
225 pp approx. 1994 [3-540-19895-4]

Vol. 199: Cohen, G.; Quadrat, J.-P. (Eds)
11th International Conference on
Analysis and Optimization of Systems –
Discrete Event Systems: Sophia-Antipolis,
June 15–16–17, 1994
648 pp. 1994 [3-540-19896-2]

Vol. 200: Yoshikawa, T.; Miyazaki, F. (Eds)
Experimental Robotics III: The 3rd
International Symposium, Kyoto, Japan,
October 28-30, 1993
624 pp. 1994 [3-540-19905-5]

Vol. 201: Kogan, J.
Robust Stability and Convexity
192 pp. 1994 [3-540-19919-5]

Vol. 202: Francis, B.A.; Tannenbaum, A.R. (Eds)
Feedback Control, Nonlinear Systems, and Complexity
288 pp. 1995 [3-540-19943-8]

Vol. 203: Popkov, Y.S.
Macrosystems Theory and its Applications: Equilibrium Models
344 pp. 1995 [3-540-19955-1]

Vol. 204: Takahashi, S.; Takahara, Y.
Logical Approach to Systems Theory
192 pp. 1995 [3-540-19956-X]

Vol. 205: Kotta, U.
Inversion Method in the Discrete-time Nonlinear Control Systems Synthesis Problems
168 pp. 1995 [3-540-19966-7]

Vol. 206: Aganovic, Z.;.Gajic, Z.
Linear Optimal Control of Bilinear Systems with Applications to Singular Perturbations and Weak Coupling
133 pp. 1995 [3-540-19976-4]

Vol. 207: Gabasov, R.; Kirillova, F.M.; Prischepova, S.V.
Optimal Feedback Control
224 pp. 1995 [3-540-19991-8]

Vol. 208: Khalil, H.K.; Chow, J.H.; Ioannou, P.A. (Eds)
Proceedings of Workshop on Advances inControl and its Applications
300 pp. 1995 [3-540-19993-4]

Vol. 209: Foias, C.; Özbay, H.; Tannenbaum, A.
Robust Control of Infinite Dimensional Systems: Frequency Domain Methods
230 pp. 1995 [3-540-19994-2]

Vol. 210: De Wilde, P.
Neural Network Models: An Analysis
164 pp. 1996 [3-540-19995-0]

Vol. 211: Gawronski, W.
Balanced Control of Flexible Structures
280 pp. 1996 [3-540-76017-2]

Vol. 212: Sanchez, A.
Formal Specification and Synthesis of Procedural Controllers for Process Systems
248 pp. 1996 [3-540-76021-0]

Vol. 213: Patra, A.; Rao, G.P.
General Hybrid Orthogonal Functions and their Applications in Systems and Control
144 pp. 1996 [3-540-76039-3]

Vol. 214: Yin, G.; Zhang, Q. (Eds)
Recent Advances in Control and Optimization of Manufacturing Systems
240 pp. 1996 [3-540-76055-5]

Vol. 215: Bonivento, C.; Marro, G.; Zanasi, R. (Eds)
Colloquium on Automatic Control
240 pp. 1996 [3-540-76060-1]

Vol. 216: Kulhavý, R.
Recursive Nonlinear Estimation: A Geometric Approach
244 pp. 1996 [3-540-76063-6]

Vol. 217: Garofalo, F.; Glielmo, L. (Eds)
Robust Control via Variable Structure and Lyapunov Techniques
336 pp. 1996 [3-540-76067-9]

Vol. 218: van der Schaft, A.
L_2 Gain and Passivity Techniques in Nonlinear Control
176 pp. 1996 [3-540-76074-1]

Vol. 219: Berger, M.-O.; Deriche, R.; Herlin, I.; Jaffré, J.; Morel, J.-M. (Eds)
ICAOS '96: 12th International Conference on Analysis and Optimization of Systems - Images, Wavelets and PDEs:
Paris, June 26-28 1996
378 pp. 1996 [3-540-76076-8]

Vol. 220: Brogliato, B.
Nonsmooth Impact Mechanics: Models, Dynamics and Control
420 pp. 1996 [3-540-76079-2]

Vol. 221: Kelkar, A.; Joshi, S.
Control of Nonlinear Multibody Flexible Space Structures
160 pp. 1996 [3-540-76093-8]

Vol. 222: Morse, A.S.
Control Using Logic-Based Switching
288 pp. 1997 [3-540-76097-0]

Vol. 223: Khatib, O.; Salisbury, J.K.
Experimental Robotics IV: The 4th
International Symposium, Stanford,
California,
June 30 - July 2, 1995
596 pp. 1997 [3-540-76133-0]

Vol. 224: Magni, J.-F.; Bennani, S.;
Terlouw, J. (Eds)
Robust Flight Control: A Design Challenge
664 pp. 1997 [3-540-76151-9]

Vol. 225: Poznyak, A.S.; Najim, K.
Learning Automata and Stochastic
Optimization
219 pp. 1997 [3-540-76154-3]

Vol. 226: Cooperman, G.; Michler, G.;
Vinck, H. (Eds)
Workshop on High Performance Computing
and Gigabit Local Area Networks
248 pp. 1997 [3-540-76169-1]

Vol. 227: Tarbouriech, S.; Garcia, G. (Eds)
Control of Uncertain Systems with Bounded
Inputs
203 pp. 1997 [3-540-76183-7]

Vol. 228: Dugard, L.; Verriest, E.I. (Eds)
Stability and Control of Time-delay Systems
344 pp. 1998 [3-540-76193-4]

Vol. 229: Laumond, J.-P. (Ed.)
Robot Motion Planning and Control
360 pp. 1998 [3-540-76219-1]

Vol. 230: Siciliano, B.; Valavanis, K.P. (Eds)
Control Problems in Robotics and Automation
328 pp. 1998 [3-540-76220-5]